SCANNING PROBE LITHOGRAPHY

T0137964

MICROSYSTEMS

Series Editor
Stephen D. Senturia
Massachusetts Institute of Technology

Editorial Board

Roger T. Howe, *University of California, Berkeley*
D. Jed Harrison, *University of Alberta*
Hiroyuki Fujita, *University of Tokyo*
Jan-Ake Schweitz, *Uppsala University*

OTHER BOOKS IN THE SERIES:

- **Methodology for the Modeling and Simulation of Microsystems**
  Bartlomiej F. Romanowicz
  Hardbound, ISBN 0-7923-8306-0, October 1998
- **Microcantilevers for Atomic Force Microscope Data Storage**
  Benjamin W. Chui
  Hardbound, ISBN 0-7923-8358-3, October 1998
- **Bringing Scanning Probe Microscopy Up to Speed**
  Stephen C. Minne, Scott R. Manalis, Calvin F. Quate
  Hardbound, ISBN 0-7923-8466-0, February 1999
- **Micromachined Ultrasound-Based Proximity Sensors**
  Mark R. Hornung, Oliver Brand
  Hardbound, ISBN 0-7923-8508-X, April 1999
- **Microfabrication in Tissue Engineering and Bioartificial Organs**
  Sangeeta Bhatia
  Hardbound, ISBN 0-7923-8566-7, August 1999
- **Microscale Heat Conduction in Integrated Circuits and Their Constituent Film.**
  Y. Sungtaek Ju, Kenneth E. Goodson
  Hardbound, ISBN 0-7923-8591-8, August 1999

SCANNING PROBE LITHOGRAPHY

by

Hyongsok T. Soh

Kathryn Wilder Guarini

Calvin F. Quate

Stanford University, Stanford, California

KLUWER ACADEMIC PUBLISHERS
Boston / Dordrecht / London

Distributors for North, Central and South America:
Kluwer Academic Publishers
101 Philip Drive
Assinippi Park
Norwell, Massachusetts 02061 USA
Telephone (781) 871-6600
Fax (781) 871-6528
E-Mail <kluwer@wkap.com>

Distributors for all other countries:
Kluwer Academic Publishers Group
Distribution Centre
Post Office Box 322
3300 AH Dordrecht, THE NETHERLANDS
Telephone 31 78 6392 392
Fax 31 78 6546 474
E-Mail <orderdept@wkap.nl>

 Electronic Services <http://www.wkap.nl>

ISBN 978-1-4419-4894-6

Library of Congress Cataloging-in-Publication Data

Soh, Hyongsok T.
  Scanning Probe Lithography / by Hyongsok T. Soh, Kathryn Wilder Guarini, Calvin F. Quate
       p.   cm. (Microsystems series)
  Includes bibliographical references and index.

  1.   Integrated circuits—Design and Construction. 2. Microlithography. 3. Scanning
       probe microscopy

TK7874 .S648 2001
621.3815/31 21

                                                                01-029782

Copyright © 2010 by Kluwer Academic Publishers.

All rights reserved. No part of this publication may be reproduced, stored in a retrieval
system or transmitted in any form or by any means, mechanical, photo-copying, recording,
or otherwise, without the prior written permission of the publisher, Kluwer Academic
Publishers, 101 Philip Drive, Assinippi Park, Norwell, Massachusetts 02061

Printed on acid-free paper.

Printed in the United States of America

Table of Contents

Table of Contents

List of Figures

List of Figures

List of Tables

Glossary

1D	one dimensional
2D	two dimensional
ADC	analog-to-digital converter
AFM	atomic force microscope
ARC	anti-reflection coating
ASIC	application specific integrated circuit
BESOI	bond and etch back silicon on insulator
BOE	buffered oxide etch
CAD	computer aided design
CAR	chemically amplified resist
CD	critical dimension
CMOS	complementary metal-oxide semiconductor
CMP	chemical mechanical polish
CMRR	common mode rejection ratio
CoO	cost-of-ownership
CVD	chemical vapor deposition
DI	deionized
DIBL	drain induced barrier lowering
DoF	depth of focus
DRAM	dynamic random access memory
DSP	digital signal processor
DUV	deep ultraviolet
EBL	electron beam lithography
EUV	extreme ultraviolet
FED	field emission display
FET	field effect transistor
FGA	forming gas anneal
FIB	focussed ion beam
HDLP	high density low pressure
HF	hydrofluoric acid
IC	integrated circuit
IPL	ion projection lithography
LFM	lateral force microscope
LOCOS	local oxidation of silicon
LPCVD	low pressure chemical vapor deposition
LSI	large scale integration
LTO	low temperature oxide

MOSFET .... metal-oxide-semiconductor field-effect transistor
MEMS .... micro electrical mechanical systems
MFS .... minimum feature size
MWNT .... multi-walled nanotube
NA .... numerical aperture
NGL .... next generation lithography
OAI .... off-axis illumination
OPC .... optical proximity correction
PAB .... post apply bake
PEB .... post exposure bake
PMMA .... polymethylmethacrylale
PREVAIL .... projection reduction exposure with variable axis immersion lenses
PSG .... phosphosilicate glass
PSM .... phase shift mask
RET .... resolution enhancement techniques
RIE .... reactive ion etching
RT .... room temperature
RTA .... rapid thermal anneal
SCALPEL . scattering with angular limitation projection electron-beam lithography
SCM .... scanning capacitance microscope
SEM .... scanning electron microscope
SET .... single electron transistor
SIA .... Semiconductor Industry Association
SIMOX .... separation by implantation of oxygen
SOG .... spin on glass
SOI .... silicon on insulator
SPL .... scanning probe lithography
SPM .... scanning probe microscope
STI .... shallow trench isolation
STM .... scanning tunnelling microscope
SWNT .... single-walled nanotube
TEM .... transmission electron microscope
ULSI .... ultra large scale integration
UHV .... ultra high vacuum
UV .... ultraviolet
VLSI .... very large scale integration
XRL .... X-ray lithography

Foreword

This book describes recent advances in the field of scanning probe lithography (SPL), a high-resolution patterning technique that uses a sharp tip in close proximity to a sample to pattern nanometer-scale features on the sample. SPL is capable of patterning sub-30-nm features with nanometer-scale alignment registration. It is a relatively simple, inexpensive, reliable method for patterning nanometer-scale features on various substrates. It has potential applications for nanometer-scale device research, for maskless semiconductor lithography, and for photomask patterning.

The early section of the book provides an introduction to scanning probes and scanning probe lithography techniques and applications. The book then describes the key methods for reliable SPL, including field-enhanced oxidation and low-energy electron resist exposure. For instance, SPL electron resist exposure exhibits a number of advantages over high-energy electron beam lithography, including a wider exposure latitude, improved linearity, and reduced proximity effects.

SPL can be used to fabricate functional electron devices such as 100 nm metal-oxide-semiconductor field-effect transistors. The device fabrication methods and transistor characteristics are presented. The SPL scanning speed constraints are evaluated, and parallel patterning using arrays of scanning probes is suggested as a solution for increasing the patterning throughput. To allow individual control of the patterning of each probe in an array, a transistor current source may be integrated onto the cantilever chip. Probe tip fabrication methods are also included, along with a discussion of carbon nanotubes as scanning probe tips.

Preface

The idea of using a microscope for *writing* as well as *reading* has been well established for decades. Some early photolithographic wafer exposure tools employed microscope objectives to delineate micron-sized features for microwave semiconductor devices. In 1960, workers in both the United States and Germany demonstrated the use of a transmission electron microscope as a writing tool to delineate deep submicron features. At about the same time, the scanning electron microscope began to be used as a means of writing information for experimental storage systems (at IBM) and for writing features for semiconductor devices (at Westinghouse and Cambridge University).

So when the scanning tunneling microscope (STM) was invented and publicized it was farily straightforward to think of the possibility of using it as a lithographic writing tool. This idea was advanced at the "3-Beams" conference in 1984 and the first results (sub-100-nm-wide gold lines) demonstrated the following year. But there were many major obstacles to be overcome before such a technique could be treated as a serious contender for even mask writing (which requires only one hundredth the writing rate needed for wafer exposure).

The atomic force microscope (AFM) introduced by Binnig, Gerber, and Quate together with variants of the STM and AFM (collectively referred to as scanning probe microscopes, SPMs) enlarged the scope of the opportunities for probe-based writing. One key advance was the ability to blank the exposing beam without losing control of the tip-to-target distance. Another key milestone (1994) was the first reported fabrication with an SPM of a working 100 nm active electron device. Since then there have been a remarkable series of advances, and SPM lithography is now regared as a serious contender for both mask writing and, eventually, wafer exposure.

The authors of this book have been key players. Calvin Quate has been involved since the beginning in the early 1980s and leads the research team that is regarded as the foremost group in this exciting field. Hyongsok Tom Soh and Kathryn Wilder Guarini have been the members of this group who, in the last few years, have brought about the aforementioned remarkable series of advances in SPM lithography. Some of these advances have been in the control of the tip which has allowed the scanning speed to be increased from μm/second to mm/second. Both non-contact and in-contact writing have been demonstrated as has controlled writing of sub-100 nm lines over large steps on the substrate surface. The engineering of a custom-designed MOSFET built into each microcantilever for individual current control is another notable achievement. Micromachined arrays of probes each with individual control have been demonstarted; this is key to raising the total coverage rate to at least mm^2/second for mask-making and to cm^2/second for wafer exposure. One of the most intriging new aspects is the use of directly-grown carbon nanotubes as robust, high-resolution emitters.

In this book the authors concisely and authoritatively describe the historical context, the relevant inventions, and the prospects for eventual manufacturing use of this exciting new technology. It is very timely.

R. F. W. Pease, 2001
Professor of Electrical Engineering
Stanford University, Stanford, California

Acknowledgments

This book is based on the Ph.D. dissertations of Hyongsok Tom Soh and Kathryn Wilder Guarini in the research group of Professor Calvin F. Quate at Stanford University's Ginzton Laboratory. This book incorporates the work of the authors and a group of talented researchers with whom they collaborated. They gratefully acknowledge the technical contributions of the other Quate Group members, especially Jesse Adams, Eugene Chow, Kenneth Crozier, Daniel Fletcher, Tom Hunt, Hae-Chang Lee, Scott Manalis, Stephen Minne, George Paloczi, Todd Sulchek, Anne Verhulst, and Goksenin Yaralioglu.

The authors also acknowlege other collaborators at Stanford University for their valuable contributions to this work: Jim McVittie, Simon Wong, Patrick Yue, Thomas Kenny, Eugene Chow, Ben Chui, Pierre Khuri-Yakub, Igal Ladabaum, Gokan Percin, James Plummer, Farid Nemati, Fabian Pease, Mark McCord, Theresa Kramer, Hongjie Dai, Jing Kong, Nathan Franklin, and Alan Cassell.

The authors are indebted to their collaborators outside of Stanford including: Abdullah Atalar of Bilkent University in Turkey; Stephen Minne and Dennis Adderton of Nano-Devices; Anthony McCarthy at Lawrence Livermore National Laboratory; Jean Fréchet, David Tully, Alex Trimble, Andy Neureuther, and Nick Rau at U.C. Berkeley; Keith Perkins and Christie Marrian at the Naval Research Laboratory; David Kyser, Bill Arnold, Bhanwar Singh, and Roger Alvis of Advanced Micro Devices (AMD); and Bing Yen at IBM Almaden Research Center.

This research was sponsored in part by grants from the Semiconductor Research Corporation (SRC), the Defense Advanced Research Projects Agency (DARPA), the Office of Naval Research (ONR), and the National Science Foundation (NSF). This work made use of the Stanford Nanofabrication Facility that is a part of the National Nanofabrication Users Network funded by the NSF. HTS and KWG gratefully acknowledge support from the Leland T. Edwards Fellowship at Stanford University. KWG is pleased to acknowledge the financial support of a Fellow-Mentor-Advisor (FMA) grant from AMD and the Karel Urbanek Fellowship at Stanford University.

Hyongsok Tom Soh (tsoh@agere.com) is now a Technical Manager at Agere Systems in Murray Hill, NJ. Kathryn Wilder Guarini (kwg@us.ibm.com) is now a Research Staff Member at IBM's T. J. Watson Research Center in Yorktown Heights, NY. Calvin F. Quate (quate@ee.stanford.edu) is Professor of Electrical Engineering and Applied Physics at Stanford University in Stanford, CA.

H. T. Soh, K. W. Guarini, and C. F. Quate
Stanford, California 2001

1   *Introduction to Scanning Probe Lithography*

Semiconductor lithography is the patterning process used to define the structures that make up integrated circuits (ICs). The semiconductor industry has historically scaled down the size of printed features on ICs because scaling both improves transistor performance and reduces the area that devices occupy. Today the patterning technology employed in manufacturing is photolithography, a process that uses ultraviolet light to define submicron-sized features in photosensitive polymers. Since photolithography is rapidly approaching fundamental resolution limitations, a new high-resolution patterning technique may be required to continue the industry's trend toward higher performance electron devices, increased packing densities, and higher density memories.

This book presents a high-resolution patterning method known as scanning probe lithography (SPL). SPL uses a sharp tip in close proximity to a sample to pattern nanometer-scale features on the sample. This chapter provides the background needed to consider whether SPL might meet the semiconductor industry's future lithography needs. In this chapter we introduce the family of instruments known as scanning probe microscopes (Section 1.1), including the recent innovations that have led to the widespread use of these tools. We also discuss the various techniques for patterning using probes (Section 1.2). The requirements and challenges of semiconductor lithography are described along with an introduction to other nanolithography techniques (Section 1.3). Finally, we provide an overview of this book (Section 1.4).

1.1 The Scanning Probe Microscope

A scanning probe microscope (SPM) is an instrument that monitors the local interaction between a sharp tip and a sample surface to acquire physical, electrical, or chemical information about the surface with high spatial resolution. Today there are many different types of SPMs used for diverse applications ranging from biological probing to surface science to semiconductor metrology.

1.1.1 The Scanning Tunneling Microscope

The scanning tunneling microscope (STM), introduced in 1982 by G. Binnig and H. Rohrer, was the first probe microscope of its kind to image with angstrom-level lateral and vertical resolution [1]. A voltage bias is applied between a sharp tip (typically an etched tungsten wire) and a sample. When the tip is in close proximity with the sample surface, electrons may tunnel across the gap between the two electrodes. The tunneling current is quite sensitive to the tip-to-sample spacing and is therefore a useful measurement of that distance. In its simplest manifestation, the STM raster scans a tip over a sample surface while monitoring the tunneling current between the tip and sample. A plot of the tunnelling current as a function of position shows the sample topography with resolution on the atomic scale. A more common configuration of the STM operates in a feedback mode in which the probe tip is moved up or down to maintain a constant tunneling current [Fig 1.1(a)]. Plotting the vertical probe movement as a function of lateral position gives a three-dimensional image of the surface. STM images can reflect both the topographic and electronic structure of the surface. The STM is traditionally operated under vacuum to stabilize tunneling currents. Vibration isolation is essential for high-resolution STM imaging, and low-temperature operation minimizes thermal noise and thermal diffusion to allow superior resolution. STM micrographs provide vivid images of topography on the scale of atoms. For their work, Binnig and Rohrer received the 1996 Nobel Prize in Physics.

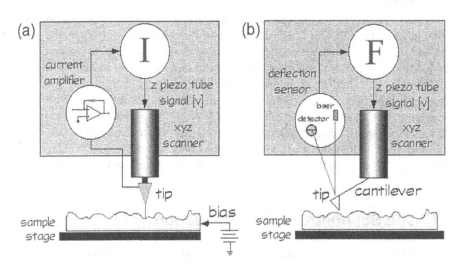

Figure 1.1: Schematic diagram of the most common configurations of the scanning probe microscope (SPM). (a) The scanning tunneling microscope (STM). (b) The atomic force microscope (AFM).

1.1.2 The Atomic Force Microscope

The primary limitation of STM is that it can only be used to image conducting substrates. The atomic force microscope (AFM) was developed to alleviate this constraint. In 1986, G. Binnig, C. F. Quate, and Ch. Gerber showed that fine sample topography can be imaged by monitoring the force between a sharp tip and sample [2]. A flexible spring-like cantilever with a sharp tip at the end is raster scanned over the sample surface. Small forces between the probe tip and sample are detected by sensing the cantilever deflection. A topographic map of the surface is generated by plotting the cantilever deflection as a function of position. Feedback control during scanning can be used to maintain a small constant contact force between the tip and sample (contact mode) [Fig 1.1(b)], a constant tip-sample spacing (in a related a.c. technique for noncontact operation [3]), or intermittent contact between the tip and sample (tapping mode [4]). All three modes have been used to generate atomic-resolution images of various conducting and insulating substrates. Most AFMs operate in air rather than vacuum, eliminating the need for costly chambers, reducing overhead time, and simplifying the measurement apparatus. The high spatial resolution and nondestructive measurement capabilities of the AFM have made it a natural choice for semiconductor metrology. Today, many commercial systems incorporate full-wafer stages, are clean-room compatible, and have extended scan ranges (> 100 μm). The AFM's ability to provide three-dimensional topographic images has been exploited by the semiconductor industry to quantify microroughness. This feature has been used, for example, to characterize chemical mechanical polish (CMP) processes and to determine the surface roughness of deposited layers [5]. Fabrication sequences may be optimized by repeating the measurements for different process conditions, thereby eliminating the need for expensive, destructive, and time-consuming cross sections. Other interesting applications of AFM semiconductor metrology include wafer inspection [6], mask inspection, overlay registration, defect imaging and analysis [7], grain-size measurements [8], etch depth measurements, and cross section imaging [9][10].

The AFM can also operate in a two-dimensional feedback mode [11]. The critical dimension AFM, or CD-AFM, scans with a boot-shaped tip. The tip shape is optimized to sense topography on vertical surfaces and can be used to profile vertical or even re-entrant sidewalls with nanometer resolution (where conventional conical or pyramidal tips fail as a result of the finite cone angle). The CD-AFM can provide feature height and width information, data inaccessible with other nondestructive measurement systems [12]. This information is critical for controlling semiconductor lithography and etch processes.

A wide array of other probe technologies has also been developed. Some scanning probe techniques provide information other than (or in addition to) topography maps [13]. For example, the lateral force microscope (LFM) detects surface friction and topography simultaneously and can differentiate between materials that may both be topographically flat [14]. The fabrication of a thermocouple on the tip enables the concurrent acquisition of topography and temperature maps. This thermal probe can be used, for instance, to image a biased MOSFET and identify the heating under the gate and the high-field "hot spot" between the gate and drain regions [15]. The Kelvin probe microscope (also known as the electrostatic force microscope) detects the contact potential difference between materials and is capable of monitoring the potential distribution in semiconductor devices [16][17]. The scanning capacitance microscope (SCM) allows access to two-dimensional carrier density data and may be used to determine the distribution of dopants in a device cross section [18].

1.1.3 Innovations Through Integration

Since its invention, the AFM has evolved from a laboratory-built instrument into a commercially-available tool. Today at least a dozen companies produce scanning probe microscopes globally as part of a $100 million industry. The evolution of the AFM into such a widely used instrument was fueled by many innovations, many of which originate from the idea of integrating various components of the AFM by means of microfabrication.

1.1.3.1 Microfabrication of Tips and Cantilevers

In the earliest configuration of the AFM, the cantilevers were hand cut from thin metal foils or formed from fine wires [19][20]. Tips for these cantilevers were prepared by attaching diamond fragments to the ends of the levers by hand, or in the case of wire cantilevers, etching the wire to a sharp point.

The AFM probe should have the following properties [21]: (1) a low force constant, (2) a high resonant frequency, (3) a high mechanical quality factor (Q), (4) a high lateral stiffness, (5) short lever length, and (6) a sharp protruding tip. It is difficult to meet the above requirements reproducibly with manually assembled AFM probes. Since the probe may need to be replaced due to damage or contamination, reliable methods were needed for fabricating large numbers of probes.

A noteworthy advancement in probe technology came when the probe (tip and the cantilever) fabrication steps were integrated through the use of microfabrication techniques by T. R. Albrecht et al. [21] in 1990. Their paper describes numerous novel methods for batch fabricating thin film SiO_2 and Si_3N_4 microcantilevers with high resonant frequencies (10-100 kHz) and with force constant small enough to detect forces of less than 10^{-8} N. Further advancements were later made in fabricating tips and cantilevers on single crystal silicon [22][23]. The batch microfabrication process provided the means to reliably produce large quantities of cantilevers and tips, which are critical, fragile, and dispensable components of the AFM.

1.1.3.2 Integrated Deflection Sensors: Piezoresistors

The second important component of the AFM is the sensor which measures the deflection of the cantilever. In the original configuration, the tunneling current between the back of the cantilever and a sharp conductive tip a few angstroms away from it was used for detecting the cantilever deflection. Displacements of 10^{-4} Angsroms could be measured when the tunneling gap was modulated. One of the problems of using tunnelling current was that the back of the cantilever gets coated with a contamination layer when the AFM is used in air. The tip must penetrate through this layer to start tunneling, which exerts a force on the cantilever that can be as large as 10^{-7} N. Also, the tunneling tip is sensitive to the topography on the backside of the cantilever, which can be a problem when the topographic variations are large. Other techniques of detecting cantilever deflection have been demonstrated, including interferometric detection [24], capacitive detection [25], and optical deflection detection [26][27].

All of the techniques mentioned above require some physical elements external to the cantilever for deflection detection. There have been many efforts to reduce the size of the detector or to integrate the detector into the cantilever [28][29]. One commercially successful approach is the piezoresistive cantilever[a] [30]. In this device, a p-type resistor is fabricated at the surface of the silicon cantilever along the <110> direction. The piezoresistive property of silicon causes its resistance to vary linearly with its deflection. Cantilevers with spring constants ranging from 1 to 10 N/m and minimum detectable deflection from 1 to 10 Angstroms were demonstrated over 10 Hz to 1 kHz.

The integration of the piezoresistive sensor allowed convenience in conditions where detector alignment is difficult, for example in ultra high vacuum (UHV) or low temperature applications. Furthermore, piezoresistive deflection sensing made it possible to form an array of scanning probes in which the deflection of each probe is individually sensed. This is the subject of discussion in Chapter 8.

1.1.3.3 Integrated Actuators and Arrays

Despite the numerous SPM modes and applications, scanning probe use is generally limited to research and development labs. Scanning probes are not typically used for routine semiconductor metrology at the production level because of their slow imaging speed. The SPM's low throughput results from the serial nature of the data acquisition scheme. All SPM techniques acquire surface maps by raster scanning a tip over every pixel in the region of interest.

---

a. For a detailed analysis of the piezoresistive cantilevers see M. Tortonese, "Force Sensor for Scanning Probe Microscopy," Ph.D. thesis, Stanford University, 1993.

One of the components that limits the scan speed of the AFM in constant force mode is the frequency response of the piezo tube scanner. To improve the bandwidth, cantilevers with integrated actuators have been developed by several groups. Some of the implementations include the use of piezoelectric films [31][32], electro-static actuation through micromachined comb-drive structures, and the use of capacitors [33]. In his thesis work, S. Minne [34] demonstrated an integrated piezoelectric actuator based on zinc oxide (ZnO) with an integrated piezoresistive sensor on a cantilever. These devices improved the speed of a typical AFM by more than an order of magnitude and allowed constant force imaging at > 1 mm/s.

Imaging throughput can be increased by scanning simultaneously with multiple tips. Arrays of AFM cantilevers with integrated tips have been fabricated using micromachining techniques [35][36]. Parallel operation requires means for controlling each tip and acquiring data from each tip simultaneously. Parallel imaging was demonstrated by operating a one dimensional array of cantilevers, each with an integrated ZnO actuator and an integrated piezoresistive sensor [35]. Higher throughput imaging with SPMs would most certainly increase the instruments' widespread use.

1.2 High-Resolution Patterning Using Scanning Probes

In addition to imaging and metrology applications, SPMs may be used to modify a sample surface. The controlled patterning of nanometer-scale features with the SPM is known as scanning probe lithography (SPL). The most common SPL techniques are introduced below. The methods differ in the patterning mechanism, achievable resolution, speed capability, and repeatability.

1.2.1 Atomic Manipulation

An SPM probe tip can be used to manufacture nanostructures by direct manipulation of nanoparticles [37], molecules [38][39], or even single atoms [40]. Particles on the surface may be moved and arranged with atomic-level precision. Figure 1.2(a) illustrates one method of atomic manipulation using a probe tip. The tip is used to "push" or "slide" a particle along the surface. Most notably, Gimzewski et al. built a nanometer-scale abacus at room temperature by controlled positioning of C_{60} molecules using an STM tip [41]. Eigler et al. used the sliding process at low temperature (4 K) to individually position 48 iron adatoms into a ring of radius 71.3 Å to form a "quantum corral" [42]. STM spectroscopy of the structure showed evidence of electron confinement within the corral. Atoms on a surface can also be manipulated by a probe using field-assisted diffusion [40]. Electrostatic and chemical forces between the tip and sample allow selective removal of individual atoms from the surface and subsequent redeposition elsewhere [40][43][44].

Atomic manipulation is a powerful tool for creating custom atomic structures and studying the physics of small-scale interactions. The high-resolution imaging capabilities of the SPM allow it to see the small structures it creates. Patterning is slow, of course, so atomic manipulation is far from a large-scale patterning technology. Nevertheless, atomic manipulation demonstrates perhaps the ultimate lithographic resolution.

1.2.2 Mechanical & Thermomechanical Patterning

An SPM probe tip can be used as a mechanical tool for patterning a sample in a process resembling "engraving" [45] or "plowing" [46]. For example, the tip might be used to displace regions (pits, lines, or craters) in a soft resist layer [Fig 1.2(b)]. The resist patterns can then be transferred into the substrate through direct etching or lift-off. A similar process has also been used to form grooves in a thin metal film [47]. The resolution of this patterning technique is limited by the tip size and the film (resist or metal) thickness. Feature dimensions below 20 nm have been created using sharp tips and thin films [48]. Repeatability is hampered by contamination and wear of the tips.

Thermomechanical patterning with a probe tip may provide a solution for high-density data storage applications [49]. The tip is kept in contact with a polycarbonate sample. For writing, the tip is heated above the glass transition

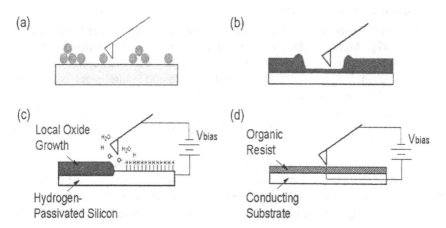

Figure 1.2: Schematic diagrams of surface modification using a scanning probe. (a) Atomic manipulation. (b) Mechanical patterning. (c) Field-enhanced oxidation of silicon. (d) Exposure of resist using electrons field-emitted from a probe tip.

temperature of the polycarbonate while exerting a small loading force on the sample. This locally melts pits (or data bits) that can be read back by measuring the cantilever deflection as the tip is scanned over a series of bits [50].

1.2.3 Local Oxidation

A voltage bias between a sharp probe tip and a sample generates an intense electric field in the vicinity of the tip. The high field can be used to locally oxidize a variety of surfaces. On silicon, this process is called *electric-field-enhanced oxidation* [Fig 1.2(c)]. Prior to patterning, a silicon sample is submersed in hydrofluoric acid (HF) to remove the native oxide and passivate the surface bonds with hydrogen. The high electric field near a negatively-biased tip locally desorbes the hydrogen passivation on the silicon surface. The field also ionizes water molecules in the air or adsorbed on the tip surface, creating negatively-charged oxygen ions. These ions are accelerated toward the sample by the electric field and react with the silicon to form an oxide film. The oxide pattern acts as a robust etch mask for subsequent pattern transfer into the silicon.

Field-enhanced oxidation of single-crystal silicon was first demonstrated by Dagata *et al.* in 1990 [51]. More recently it was discovered that amorphous silicon can be similarly oxidized [52]. Amorphous silicon is a reasonably general resist material since it can be deposited on almost any surface at low temperature. Oxide patterns can be written on amorphous silicon, transferred through wet or dry etching into the amorphous silicon layer, and subsequently transferred into the underlying film [53]. The best reported resolution of SPM-induced oxide patterns on silicon is about 10 nm [54][55].

In a related process called *anodization*, a metal sample such as titanium [56], tantalum [57], chromium [58], or aluminum [59] can be locally oxidized. The high field at the tip causes a chemical reaction between the metal and adsorbed water on the surface, resulting in local oxide growth. Anodization pattern size is typically limited by the grain size and/or surface roughness of the metal [60]. Matsumoto *et al.* fabricated a single electron transistor (SET) using the STM to create 15-nm-wide oxide patterns in a 3-nm-thick titanium film [61]. The small island allowed the SET to operate at room temperature.

These oxidation processes are powerful because of the fine resolution and the resistant oxide etch mask that is created. Patterning is slow, however, limited by the reaction rate at the surface. Writing speeds are typically limited to below 10 μm/s, making patterning of large areas in reasonable times unmanageable.[a] Local oxida-

---

a. Snow used a short pulse to oxidize silicon and from the length of the pulse inferred a maximum writing speed of 1 mm/s [54]. From our experience, however, oxidation does not reliably occur at scan speeds in excess of about 10 μm/s.

---

tion of silicon also results in significant tip wear, resulting in degraded pattern dimensions [62]. Finally, reproducibility of oxide patterns is poor because of the dependence of the pattern size on tip shape, humidity level, and surface roughness.

1.2.4 Electron Exposure of Resist

Another scheme for performing lithography with scanning probes involves the electron exposure of a resist material, such as an organic polymer or a monolayer resist. The concentrated high electric field at a biased probe tip enables the field emission of electrons from the tip. If a sample is coated with a thin resist, the emitted electrons traverse the resist [Fig 1.2(d)]. The resist absorbs energy from the electron radiation, which induces chemical changes in the resist. For organic polymer resists, the radiation scissions bonds (positive resist) or crosslinks molecules (negative resist). As a result, when the resist is submersed in a special solvent (the developer), only the irradiated areas dissolve (for positive resists; opposite for negative resists). The resist pattern can then be transferred into the substrate using selective chemical etching or dry etching. Electron exposure using a scanning probe was first demonstrated by McCord and Pease in the mid-1980s when they showed that an STM could be used to pattern organic and inorganic resists [63].

Organic polymer resists are an attractive option for SPL because many possess a low threshold voltage, high sensitivity, sub-100-nm resolution, and good dry etch resistance. Moreover, these resists can be easily coated on virtually any substrate, and many organic polymers are well characterized from their use in photolithography or electron beam lithography. It is also possible to pattern at higher speeds with this method than with local oxidation. In one example, siloxene (spin on glass, SOG) was patterned with the tip scanning above 3 mm/s [64]. Finally, the polymer surface is soft and pliable, which significantly reduces the tip wear.

1.3 Semiconductor Lithography

Today microprocessors contain about 6 million transistors in each square centimeter. By 2011 that number is expected to grow to 180 million. It is predicted that the microprocessor on-chip frequency will jump from 600 MHz to 1.8 GHz over the next decade. DRAM memory cells that today hold 1 Gbit of data will contain as many as 256 Gbits in 2011. These performance improvements are all projected to be met with decreased cost-per-transistor and cost-per-bit (Table 1.1) [65]. Lithography is the key technology that will enable (or limit) this rapid growth.

IC fabrication relies on a series of material deposition, patterning, and etching steps. Lithography costs account for more than 35% of the total chip manufacturing costs [65]. Microprocessors today incorporate more than 20 levels of lithography. The smallest pattern printed on an IC (known as the critical dimension, or CD) has decreased exponen- tially over time. This phenomenon, known as Moore's law [66], has evolved from a historical trend into an impetus that drives the semiconductor industry toward faster microprocessors and higher density memories. Today the microprocessor CD is about 180 nm. This dimension must be reduced to 50 nm by 2011 to meet future microprocessor and DRAM performance goals (Table 1.1).

Device scaling poses a number of challenges for semiconductor lithography. First is the ability to reduce the minimum feature size printed by a patterning system (*resolution*). As IC dimensions shrink, more demands will be placed on lithography process control. The accuracy and repeatability of the feature dimension is critical for device yield and performance (*process latitude*). Pattern placement is important for any multi-level lithography scheme (*overlay registration*). Economic factors dictate that the small patterns be created quickly and efficiently (*throughput*). Finally, there must be a reliable method for faithfully transferring a narrow feature from the sacrificial resist layer into the underlying film (*pattern transfer*).

Year of First Shipment:	1999	2002	2005	2008	2011
Dense Line CD (nm)	180	130	100	70	50
Isolated Line CD (nm)	140	100	70	50	35
Microprocessor On-Chip Frequency[a] (MHz)	600	800	1100	1400	1800
Transistors/cm^2	6.2M	18M	39M	84M	180M
Cost/Transistor (microcents)	1735	580	255	110	49
DRAM Bits	1G	4G	16G	64G	256G
Bits/cm^2	270M	770M	2.2B	6.1B	17B
Cost/Bit (microcents)	40	15	5.3	1.9	0.66
Gate CD Control[b] (nm)	14	10	7	5	4
Overlay Registration[c] (nm)	65	45	35	25	20
Depth of Focus[d] (μm)	0.7	0.6	0.5	0.5	0.5
Wafer Diameter (mm)	200	300	300	450	450

Table 1.1:   Future IC performance and critical level lithography requirements as projected by the SIA International Technology Roadmap for Semiconductors [65].

a. On-chip, across-chip clock, cost-performance.
b. 3σ, post-etch tolerance.
c. Mean + 3σ.
d. Usable @ full field ± 10% exposure.

1.3.1 Photolithography

Projection photolithography is the dominant patterning technique used in semiconductor manufacturing today. A photolithography system is composed of four primary elements:

(1) a high-intensity light source,

(2) a chrome-on-glass patterned mask plate,

(3) optical elements (including filters that select the specific exposure wavelength and lenses that reduce the size of features from the mask to the wafer), and

(4) a wafer coated with a photosensitive polymer film (known as the photoresist).

The mask is illuminated by ultraviolet (UV) light, which passes through the glass regions and is absorbed by the chrome. The transmitted light "exposes" the photoresist, initiating chemical changes in the polymer that result in differential solubility between the exposed and unexposed regions. The wafer is then submersed in the developer solution, leaving a three-dimensional relief image in the resist. The resist acts as an etch mask for subsequent pattern transfer into the underlying film.

The mask is patterned with either direct write electron beam lithography (EBL) or a laser writer. It contains the patterns to be printed in one exposure field (typically a few cm^2). A step-and-repeat photolithography tool (known as an optical stepper) repeatedly projects the mask pattern onto the wafer until the full wafer is covered. A step-and-scan system uses a field of view smaller than the mask pattern. It scans the mask and wafer simultaneously during exposure, then steps to the next field position on the wafer and repeats the scan. Modern photolithography systems pattern more than 40 200-mm-diameter wafers per hour. This is equivalent to a transfer rate of nearly 30 billion pixels per second from the mask to the wafer [67].

1.3.1.1 Photolithography Limitations

The resolution of photolithography is fundamentally limited by optical diffraction. The minimum feature size (MFS) that can be printed depends on the wavelength (λ) of the exposing light and the numerical aperture (NA) of the exposure system:

$$MFS = k_1 \frac{\lambda}{NA} \qquad (1.1)$$

NA is the sine of the convergence angle of the lens and k_1 is typically 0.5-0.8, depending mainly on the resist technology. To improve the resolution, we can reduce λ and/or increase NA. Both have the deleterious effect of decreasing the system depth of focus (DoF), which goes as:

$$DoF = \pm k_2 \frac{\lambda}{NA^2} \qquad (1.2)$$

where k_2 is typically 0.1-0.5 [68].

1.3.1.2   Resolution Enhancement Techniques

During the past decade the semiconductor industry has reduced the exposure wavelength from 435 nm (g-line) to 365 nm (i-line) to 248 nm (deep ultraviolet, DUV) in an effort to improve patterning resolution. DUV lithography, which uses 248 nm light generated by a krypton fluoride (KrF) laser source, has been used in semiconductor manufacturing since 1998. The reduced DoF of DUV lithography tools has forced the industry to move to thinner resists and to planarize the sample surface using techniques such as CMP. Interference of light has necessitated the use of anti-reflective coatings (ARCs) for all critical patterning levels.

The exposure wavelength cannot be reduced indefinitely because the optical elements in an exposure tool that are meant to transmit the light, such as the lenses and masks, become absorbing at short wavelengths. Exposure tools with 193 nm radiation are currently under development. The 193 nm light generated by an argon fluoride (ArF) laser source might be acceptable for reaching the 100 nm node (in 2005, see Table 1.1). Optical lithography may be extended to λ=157 nm, but would require new lens materials (calcium fluoride or fused silica) and might be used down to the 70 nm node (in 2008, see Table 1.1). There remain a number of engineering challenges for 157 nm lithography [69]. No known lens materials are suitable for λ<157 nm.

A variety of resolution enhancement techniques (RETs) have allowed each wavelength to be pushed farther than the diffraction-limited resolution described above. Optical proximity correction (OPC) is used in production today to improve pattern fidelity. OPC modifies the mask pattern to account for optical diffraction and produce the desired feature shape and size when printed on the wafer. Phase shift masks (PSMs) may be used in place of standard binary chrome-on-glass optical masks. PSMs introduce destructive interference effects that enhance the image contrast. Off-axis illumination (OAI) can improve both resolution and DoF.

1.3.2 Beyond Optical: Technologies for Next Generation Lithography

Several alternative lithographic technologies have been proposed for use in semiconductor manufacturing for printing features beyond the resolution limits of optical lithography. Some next generation lithography (NGL) techniques use shorter wavelength photons (extreme ultraviolet radiation or x-rays) to minimize diffraction effects. Other schemes use charged particle beams (electrons or ions) for patterning or no beams at all (contact printing). Scanning probe lithography is proposed here as an alternative. An NGL solution must meet the future lithography requirements (including resolution, overlay, and DoF), be cost-competitive (including tool costs and achievable throughput), be extendable to future generations, and be compatible with the existing semiconductor manufacturing infrastructure. We provide here a brief introduction to the other promising NGL contenders.

1.3.2.3   Extreme Ultraviolet Lithography

Extreme ultraviolet (EUV) lithography exposes resist with 13.4 nm wavelength photons (sometimes called "soft x-rays") generated by a laser plasma source. At this short wavelength, most materials are absorbing. Therefore EUV employs reflective optics in its reduction stepper systems. The lenses and masks are replaced by reflective multilayers. A 40-pair multilayer stack (such as Mo-Si on a 6.7 nm period) can achieve about 60% reflectivity. The mask contains a patterned absorber (such as Al) on top of the multilayer. EUV mask fabrication and inspection pose significant challenges. Since polymers are highly absorbing at 13.4 nm, thin resists or top surface imaging (TSI) resist technology must be adopted. An EUV system must be run under vacuum since air absorbs 13.4 nm radiation. EUV lithography has been used to pattern the gates of 130 nm MOSFETs [70]. It is believed that EUV resolution is extendable to below 30 nm [71].

1.3.2.4   X-Ray Lithography

X-ray lithography (XRL) exposes resist using 7-12 Å wavelength x-rays generated by a synchrotron source. Many x-ray steppers can be run off of a single electron storage ring. The x-rays are directed through proximity shadow masks onto a resist-coated sample. Since x-rays cannot be focussed, XRL is a 1:1 printing system, which places severe restrictions on mask fabrication. X-ray masks are thin membranes (typically 2-µm-thick silicon carbide) that support an absorber pattern (usually a heavy metal) [72]. The gap between the wafer and mask is usually 10-20 µm. XRL can pattern narrow features in thick resist with a respectable throughput of over 10 200-mm-diameter wafers per hour. Resolution has been demonstrated down to 70 nm. A number of devices have been fabricated using XRL, including complementary metal-oxide-semiconductor (CMOS) test circuits with features sizes of 100 nm [73].

1.3.2.5   Electron Beam Lithography

Direct-write electron beam lithography (EBL) scans a finely focussed beam of high energy (typically 10–100 keV) electrons across the sample under computer control. The de Broglie wavelength of an EBL electron is on the order of 1 Å, so diffraction effects are negligible. Features as small as 10 nm have been written in resist with EBL [74]. Proximity effects resulting from backscattered electrons hinder line width control. Coulomb interaction effects cause beam blur at high current levels and limit throughput. Direct-write EBL is the primary method for patterning optical, EUV, and x-ray masks, but it is not considered a viable solution for high-volume lithography applications because of the low throughput. There is some effort to build EBL systems with multiple electron beams for increased-throughput parallel lithography [75][76][77].

Projection EBL is a mask-based approach that offers improved throughput while retaining many of the advantages of an electron exposure tool. Projection EBL employs scattering masks consisting of a thin membrane (typically 100-nm-thick silicon nitride), a patterned tungsten scattering layer, and support struts. Exposure contrast is achieved through the difference in the electron scattering characteristics between the membrane and tungsten. Projection EBL is currently under investigation by Lucent Technologies and IBM. Lucent's scheme is dubbed "scattering with angular limitation projection electron beam lithography" (SCAL-PEL) [78] and IBM's system is known as "projection reduction exposure with variable axis immersion lenses" (PREVAIL) [79]. Projection EBL is expected to achieve a throughput of about 24 300-mm-diameter wafers per hour and be extendable to 50 nm feature sizes.

1.3.2.6 Ion Projection Lithography

Ion projection lithography (IPL) irradiates a stencil mask with a uniform beam of light ions such as H^+ or He^+. The transmitted image is demagnified and projected onto a resist-coated substrate. Since ions scatter very little in the resist, the line width is not a strong function of dose. IPL has a large DoF, negligible proximity effects, and demonstrated resolution down to 70 nm (equal line/space) [80]. IPL steppers must run under vacuum conditions. Space charge limitations can result in pattern blur and degraded placement accuracy. IPL steppers are expected to have a very high throughput of over 70 200-mm-diameter wafers per hour [81].

1.3.2.7 Contact Printing

Contact printing is a high-speed, low-cost patterning method. In contrast to the other schemes described above, contact printing involves no "exposure" by energetic beams and therefore is not limited by diffraction or scattering. Instead, it involves deforming resist by embossing with a master mold. There are a variety of implementations of this concept, all of which use a three-dimensional master to transfer large-area patterns onto the substrate. Achieving accurate overlay registration is the primary challenge for contact printing.

Chou's "nanoimprint lithography" process involves physically deforming a resist layer on the surface while controlling the mold pressure and temperature. After imprinting, the mold is removed. Reactive ion etching (RIE) is used to remove any remaining polymer in the compressed areas. Features down to 25 nm dimensions have been patterned in this way [82]. In a "step-and-flash" technique used by Willson, a master quartz mold is aligned in close proximity to the substrate. A UV-curable solution fills the voids in the mold by capillary action. The mold is irradiated to cure the solution and then the mold is separated from the patterned film. The pattern can then be etched into the substrate. This technique can achieve sub-60-nm resolution [83].

1.4 Book Overview

The remainder of this book presents the capabilities of scanning probe lithography (SPL) for reliable high-resolution patterning. We focus in particular on the ability of SPL to meet the future lithography requirements of the semiconductor industry. Chapter 2 presents SPL using electric-field-enhanced oxidation of silicon. We show that this SPL method can be used to pattern the gate of a 100-nm nMOSFET.

Chapter 3 presents our methods for reliably patterning nanometer-scale features in organic resist using a scanning probe. We incorporated real-time electron exposure dose control to ensure uniform patterning and to set the feature size. We present both a contact mode system, in which the tip is scanned in contact with the resist surface, and a noncontact mode system, which reliably patterns without risk of damage to the tip or sample. We discuss the exposure characteristics and demonstrate pattern transfer through direct etching and lift-off. We also show simulation results of electron field emission from a tip and resulting beam spreading.

In Chapter 4 we investigate SPL linewidth control. We compare resist exposure by low-energy SPL electrons to exposure by high-energy EBL electrons. We show that SPL has a wider exposure latitude, improved linearity, and reduced proximity effects as compared to EBL. We provide an analysis of the distribution of absorbed energy density in the resist by EBL and SPL and discuss the exposure mechanisms of high-energy and low-energy electrons.

Chapter 5 addressed SPL's ability to meet the semiconductor industry's critical dimension lithography requirements. We describe the fabrication of short-channel transistors using SPL for gate level patterning. This demonstrates that SPL is compatible with standard semiconductor processing techniques. The gate level lithography process required uniform patterning of nanometer-scale features over sample topography and accurate overlay registration.

Chapter 6 discusses the throughput capabilities and limitations of SPL. Because of the serial nature of this direct write lithography technique, the throughput is typically lower than that of other patterning systems. We demonstrate reliable high-speed patterning of organic resists.

Chapter 7 presents an integrated transistor current source for on-chip control of the electron exposure dose during SPL. The integrated current source eliminates the need for external circuitry and may provide a solution for independently-controlled SPL with multiple probes. We discuss the design and fabrication of the integrated current source and demonstrate that it can be used as the sole current-control electronics for reliable SPL.

Chapter 8 discusses scanning probe arrays for high speed parallel lithography. We demonstrate parallel, current-controlled patterning with two probes and the extension to two-dimensional arrays. We discuss the challenges for extending SPL to many tips in parallel. We evaluate the scan speeds and probe densities required to make SPL competitive with other patterning systems.

Finally, Chapter 9 presents scanning probe tips, including the requirements, wear properties, and tip fabrication methods for scanning probe lithography. The utility of carbon nanotubes as scanning probe tips is discussed.

1.5 References

[1]   G. Binnig and H. Rohrer, "Scanning tunneling microscopy," Helvetica Physica Acta **55**, 726-735 (1982).

[2]   G. Binnig, C. F. Quate, and Ch. Gerber, "Atomic force microscope," Phys. Rev. Lett. **56**, 930-933 (1986).

[3]   P. C. D. Hobbs, Y. Martin, C. C. Williams, and H. K. Wickramasinghe, "Atomic force microscope: implementations," Proc. SPIE **897**, 26-30 (1988).

[4]   C. B. Prater and Y. E. Strausser, "Tapping mode atomic force microscopy: Applications to semiconductors," Proc. 5th International Conference on Defect Recognition and Image Processing in Semiconductors and Devices, 69-72 (1993).

[5]   D. Pramanik, M. Weling, and L. Zhou, "Using AFM to develop sub-μm multilevel metallization processes," Solid State Technology **37**, 79-86 (1994).

[6]   B. Burggraaf, "Pursuing advanced metrology solutions," Semiconductor Int. **17**, 62-64 (1994).

[7]   J. Li, S. Xiao, A. Zhao, and D. Li, "Inspecting of the microprofile and defects of optical surfaces using the atomic force microscope," Proc. SPIE **3422**, 270-273 (1998).

[8]   M. R. Rodgers, M. A. Wendman, and F. D. Yashar, "Application of the atomic force microscope to integrated circuit failure analysis," Microelectron. Reliab. **33**, 1947-1956 (1993).

References

[9]   G. Neubauer and M. L. A. Dass, "Imaging VLSI cross sections by atomic force microscopy," Proc. IEEE Reliability Physics, 299-303 (1992).

[10]   K. Wilder, C. F. Quate, B. Singh, and W. H. Arnold, "Cross sectional imaging and metrology using the atomic force microscope," J. Vac. Sci. Technol. B **14**, 4004-4008 (1996).

[11]   Y. Martin and H. K. Wickramasinghe, "Method for imaging sidewalls by atomic force microscopy," Appl. Phys. Lett. **64**, 2498-2500 (1994).

[12]   K. Wilder, B. Singh, and W. H. Arnold, "Novel in-line applications of atomic force microscopy," Sol. State Technol. **39**, 5, 109-116 (1996).

[13]   H. K. Wickramasinghe, "Scanned-probe microscopes," Scientific American **261**, 74-81 (1989).

[14]   M. R. Rodgers, M. A. Wendman, and F. D. Yashar, "Application of the atomic force microscope to integrated circuit failure analysis," Microelectron. Reliab. **33**, 1947-1956 (1993).

[15]   A. Majumdar, J. P. Carrejo, and J. Lai, "Thermal imaging using the atomic force microscope," Appl. Phys. Lett. **62**, 2501-2503 (1993).

[16]   N. Nonnenmacher, M. P. O'Boyle, and H. K. Wickramasinghe, "Kelvin probe force microscopy," Appl. Phys. Lett. **68**, 2921-2923 (1991).

[17]   O. Vatel and M. Tanimoto, "Kelvin probe force microscopy for potential distribution measurements of semiconductor devices," J. Appl. Phys. **77**, 2358-2368 (1995).

[18]   G. Neubauer, A. Erickson, C. C. Williams, J. J. Kopanski, M. Rodgers, and D. Adderton, "Two-dimensional scanning capacitance microscopy measurements of cross-sectioned very large scale integrated test structures," J. Vac. Sci. Technol. B **14**, 426-432 (1996).

[19]   Y. Martin, C. C. Williams, and H. K. Wickramasinghe, "Atomic force microscope-force mapping and profiling on a sub 100-A scale," J. Appl. Phys. **10**, 4723 (1987).

[20]   O. Marti, B. Drake, and P. K. Hansma, "Atomic force microscopy of liquid-covered surfaces: Atomic resolution images," Appl. Phys. Lett. **7**, 484 (1987).

[21]   T. R. Albrecht, S. Akamine, T. E. Carver, and C. F. Quate, "Microfabrication of cantilever styli for the atomic force microscope," J. Vac. Sci Technol. A. **8**, 3386 (1990).

[22]   M. M. Farooqui, A. G. R. Evans, M. Stedman, and J. Haycocks, "Micromachined silicon sensors for atomic force microscopy," Nanotechnology **3**, 91 (1992)

[23] S. Akamine, R. C. Barrett, and C F. Quate, "Improved atomic force microscope images using microcantilevers with sharp tips," Appl. Phys. Lett. **57**, 316 (1990).

[24] R. Erlandsson, G. M. McClelland, C. M. Mate, and S. Chiang, "Atomic force microscopy using optical interferometry," J. Vac. Sci Technol. **A6**, 266 (1988).

[25] T. Goddenhenrich, H. Lemke, U. Hartmann, and C. Heiden, "Force microscope with capacitive displacement detection," J. Vac. Sci. Technol. **A8**, 383 (1990).

[26] G. Meyer and N. M. Amer, "Novel optical approach to atomic force microscopy," Appl. Phys. Lett. **53**, 1045 (1988).

[27] S. Alexander, L. Hellemans, O. Marti, J. Schneir, V. Elings, P. K. Hansma, M. Longmire, and J. Gurley, "An atomic-resolution atomic-force microscope implemented using an optical lever," J. Appl. Phys. 65, 164-167, (1989).

[28] D. Sarid, P. Pax, L. Yi, S. Howells, M. Gallagher, T. Chen, V. Elings, and D. Bocek, "Improved atomic force microscope using a laser diode interferometer," Rev. Sci. Instrum. **63**, 3905 (1992).

[29] J. Brugger, R. A. Buser, and N. F. de Rooij, "Micromachined atomic force microprobe with integrated capacitive read-out," J. Micromech. Microeng. **2**, 218 (1992).

[30] M. Tortonese, H. Yamada, R. C. Barrett, and C .F. Quate, "Atomic force microscopy using a piezoresistive cantilever," Proceedings of Transducers '91, IEEE publication 91CH2817-5, 448 (1991).

[31] S. Akamine, T. R. Albrecht, M. J. Zdeblick, and C. F. Quate, "Microfabricated scanning tunneling microscope," IEEE Electron Device Letters **10**, 490 (1989).

[32] T. Fujii, S. Watanabe, M. Suzuki, and T. Fujiu, "Application of lead zirconate titanate thin film displacement sensors for the atomic force microscope," J. Vac, Sci. Technol. B **13**, 1119 (1995).

[33] N .C. MacDonald, "Single crystal silicon nanomechanisms for scanned-probe device arrays," Technical Digest. IEEE Solid-State Sensor and Actuator Workshop (Cat. No.92TH0403-X), 1, (1992).

[34] S. C. Minne S. R. Manalis, and C. F. Quate, "Parallel atomic force microscopy using cantilevers with integrated piezoresistive sensors and integrated piezoelectric actuators," Appl. Phys. Lett. **67**, 3918 (1995).

[35] S. C. Minne, G. Yaralioglu, S. R. Manalis, J. D. Adams, J. Zesch, A. Atalar, and C. F. Quate, "Automated parallel high-speed atomic force microscopy," Appl. Phys. Lett. **72**, 2340-2342 (1998).

[36]   M. Lutwyche, C. Andreoli, G. Binnig, J. Brugger, U. Drechsler, W. Haberle, H. Rohrer, H. Rothuizen, P. Vettiger, G. Yaralioglu, and C. Quate, "5×5 2D AFM cantilever arrays: A first step towards a Terabit storage device," Sens. Actuators **A73**, 89 (1999).

[37]   C. Baur, A. Bugacov, B. E. Koel, A. Madhukar, N. Montoya, T. R. Ramachandran, A. A. G. Requicha, R. Resch, and P. Will, "Nanoparticle manipulation by mechanical pushing: Underlying phenomena and real-time monitoring," Nanotechnology **9**, 360-364 (1998).

[38]   P. H. Beton, A. W. Dunn, and P. Moriarty, "Manipulation of C_{60} molecules on a Si surface," Appl. Phys. Lett. **67**, 1075-1077 (1995).

[39]   T. A. Jung, R. R. Schlittler, J. K. Gimzewski, H. Tang, and C. Joachim, "Controlled room-temperature positioning of individual molecules: Molecular flexure and motion," Science **271**, 181-184 (1996).

[40]   J. A. Stroscio and D. M. Eigler, "Atomic and molecular manipulation with the scanning tunneling microscope," Science **254**, 1319-1326 (1991).

[41]   M. T. Cuberes, R. R. Schlitter, and J. K. Gimzewski, "Room-temperature repositioning of individual C_{60} molecules at Cu steps: Operation of a molecular counting device," Appl. Phys. Lett. **69**, 3016-3018 (1996).

[42]   M. F. Crommie, C. P. Lutz, and D. M. Eigler, "Confinement of electrons to quantum corrals on a metal surface," Science **262**, 218-220 (1993).

[43]   I.-W. Lyo and P. Avouris, "Field-induced nanometer- to atomic-scale manipulation of silicon surface with STM," Science **253**, 173-176 (1991).

[44]   M. A. McCord and R. F. W. Pease, "A scanning tunneling microscope for surface modification," J. Phys. Colloq. **47**, 485-491 (1986).

[45]   V. Bouchiat and D. Esteve, "Lift-off lithography using an atomic force microscope," Appl. Phys. Lett. **69**, 3098-3100 (1996).

[46]   L. L. Sohn and R. L. Willett, "Fabrication of nanostructures using atomic-force-microscope-based lithography," Appl. Phys. Lett. **67**, 1552-1554 (1995).

[47]   S. Hu, S. Altmeyer, A. Hamidi, B. Spangenberg, and H. Kurz, "A novel approach to atomic force lithography," J. Vac. Sci. Technol. B **16**, 1983-1986 (1998).

[48]   S. Hu, A. Hamidi, S. Altmeyer, T. Koster, B. Spangenberg, and H. Kurz, "Fabrication of silicon and metal nanowires and dots using mechanical atomic force lithography," J. Vac. Sci. Technol. B **16**, 2822-2824 (1998).

[49]   H. J. Mamin, B. D. Terris, L. S. Fan, S. Hoen, R. C. Barrett, and D. Rugar, "High-density data storage using proximal probe techniques," IBM J. of Res. and Dev. **39**, 681-699 (1995).

[50]   H. J. Mamin and D. Rugar, "Thermomechanical writing with an atomic force microscope tip," Appl. Phys. Lett. **61**, 1003-1005 (1992).

[51]   J. A. Dagata, J. Schneir, H. H. Harary, C. J. Evans, M. T. Postek, and J. Bennett, "Modification of hydrogen-passivated silicon by a scanning tunneling microscope operating in air," Appl. Phys. Lett. **56**, 2001-2003 (1990).

[52]   N. Kramer, J. Jorritsma, H. Birk, and C. Schönenberger, "Nanometer lithography on silicon and hydrogenated amorphous silicon with low energy electrons," J. Vac. Sci. Technol. B **13**, 805-811 (1995).

[53]   S. C. Minne, P. Flueckiger, H. T. Soh, and C. F. Quate, "Atomic force microscope lithography using amorphous silicon as a resist and advances in parallel operation," J. Vac. Sci. Technol. B **13**, 1380-1385 (1994).

[54]   E. S. Snow and P. M. Campbell, "Fabrication of Si nanostructures with an atomic force microscope," Appl. Phys. Lett. **64**, 1932-1934 (1994).

[55]   H. Dai, N. Franklin, and J. Han, "Exploiting the properties of carbon nanotubes for nanolithography," Appl. Phys. Lett. **73**, 1508-1510 (1998).

[56]   H. Sugimura, T. Uchida, N. Kitamura, and H. Masuhara, "Tip-induced anodization of titanium surfaces by scanning tunneling microscopy: A humidity effect on nanolithography," Appl. Phys. Lett. **63**, 1288-1290 (1993).

[57]   T. Thundat, L. A. Nagahara, P. I. Oden, S. M. Lindsay, M. A. George, and W. S. Glaunsinger, "Modification of tantalum surfaces by scanning tunneling microscopy in an electrochemical cell," J. Vac. Sci. Technol. A **8**, 3537-3541 (1990).

[58]   H. J. Song, M. J. Rack, K. Abugharbieh, S. Y. Lee, V. Khan, D. K. Ferry, and D. R. Allee, "25 nm chromium oxide lines formed by scanning tunneling lithography in air," J. Vac. Sci. Technol. B **12**, 3720-3724 (1994).

[59]   E. S. Snow, D. Park, and P. M. Campbell, "Single-atom point contact devices fabricated with an atomic force microscope," Appl. Phys. Lett. **69**, 269-271 (1996).

[60]   H. Sugimura, T. Uchida, N. Kitamura, and H. Mauhara, "Scanning tunneling microscope tip-induced anodization of titanium: characterization of the modified surface and application to the metal resist process for nanolithography," J. Vac. Sci. Technol. B **12**, 2884-2888 (1994).

[61]   K. Matsumoto, M. Ishii, K. Segawa, Y. Oka, B. J. Vartanian, and J. S. Harris, "Room temperature operation of a single electron transistor made by the scanning tunneling microscope nanoxidation process for the TiO_x/Ti system," Appl. Phys. Lett. **68**, 34-36 (1996).

[62]   S. C. Minne, S. R. Manalis, A. Atalar, and C. F. Quate, "Independent parallel lithography using the atomic force microscope," J. Vac. Sci. Technol. B **14**, 2456-2461 (1996).

[63]   M. A. McCord, "Lithography with the scanning tunneling microscope," Ph.D. Thesis, Stanford University (1987).

[64]   S. W. Park, H. T. Soh, C. F. Quate, and S.-I. Park, "Nanometer scale lithography at high scanning speeds with the atomic force microscope using spin on glass," Appl. Phys. Lett. **67**, 2415-2417 (1995).

[65]   *International Technology Roadmap for Semiconductors* (San Jose: Semiconductor Industry Association, 1997). Data also reflects 1998 update to the roadmap.

[66]   G. E. Moore, "Progress in digital integrated electronics," Proc. IEDM, 11-13 (1975).

[67]   H. Levinson, "How far will optics take us?" presented at Stanford University 5/10/99.

[68]   L. F. Thompson, C. G. Willson, and M. J. Bowden, *Introduction to Microlithography* (Washington, DC: American Chemical Society, 1994).

[69]   M. Rothschild, "157 nm: The deepest deep-UV yet," presented at the 43rd International Conference on Electron, Ion, and Photon Beam Technology and Nanofabrication, Marco Island, FL, 1-4 June 1999.

[70]   D. A. Tichenor, G. D. Kubiak, and R. H. Stulen, "Extreme ultraviolet lithography for circuit fabrication at 0.1 µm feature size," Proc. SPIE **2523**, 23-28 (1995).

[71]   C. W. Gwyn, "Extreme ultraviolet lithography," J. Vac. Sci. Technol. B **16**, 3142-3149 (1998).

[72]   J. P. Silverman, "X-ray lithography: Status, challenges, and outlook for 0.13 µm," J. Vac. Sci. Technol. B **15**, 2117-2124 (1997).

[73]   S. J. Wind, Y. Taur, Y. H. Lee, R. G. Viswanathan, J. J. Bucchignano, A. T. Pomerene, R. M. Sicina, K. R. Milkove, J. W. Stiebritz, R. A. Roy, C. K. Hu, M. P. Manny, S. Cohen, W. Chen, "Lithography and fabrication processes for sub-100-nm scale complementary metal-oxide-semiconductor devices and circuits," J. Vac. Sci. Technol. B **13**, 2688-2695 (1995).

[74]   A. Broers, in *Nanostructure Physics and Fabrication,* edited by M. A. Reed and
        W. P. Kirk (Academic Press, San Diego, CA, 1989), p. 421.

[75]   T. H. P. Chang, M. G. R. Thomson, E. Kratschner, H. S. Kim, M. L. Yu, K. Y.
        Lee, S. A. Richtson, and B. W. Hussey, "Electron beam microcolumns for
        lithography and related applications," J. Vac. Sci. Technol. B **14**, 3774-3781
        (1996).

[76]   A. W. Baum, J. E. Schneider, R. F. W. Pease, M. A. McCord, W. E. Spicer, K. A.
        Costello, V. W. Aebi, "Semiconductor on glass photocathodes for high throughput
        maskless electron beam lithography," J. Vac. Sci. Technol. B **15**, 2707-2712
        (1997).

[77]   G. Winograd, L. Han, M. A. McCord, R. F. W. Pease, and V. Krishnamurthi,
        "Multiplexed blanker array for parallel electron beam lithography," J. Vac. Sci.
        Technol. B **16**, 3175-3176 (1998).

[78]   J. A. Liddle, *et al.,* "The scattering with angular limitation in projection electron-
        beam lithography (SCALPEL) system," Jpn. J. Appl. Phys. 1 **35**, 6663-6671
        (1995).

[79]   H. C. Pfeiffer, *et al.,* "Projection reduction exposure with variable axis immersion
        lenses: Next generation lithography," J. Vac. Sci. Technol. B **17**, 2840-2846
        (1999).

[80]   J. Melngailis, A. A. Mondelli, I. L. Berry, and R. Mohondro, "A review of ion
        projection lithography," J. Vac. Sci. Technol. B **16**, 927-957 (1998).

[81]   I. L. Berry, "Economic and technical case for ion projection lithography," J. Vac.
        Sci. Technol. B **16**, 2444-2448 (1998).

[82]   S. Y. Chou, P. R. Krauss, and P. J. Renstrom, "Nanoimprint lithography," J. Vac.
        Sci. Technol. B **14**, 4129 (1996).

[83]   M. Colburn, S. Johnson, M. Stewart, S. Damle, T. Bailey, B. Choi, M. Wedlake, T.
        Michaelson, S.V. Sreenivasan, J. Ekerdt, and C. G. Willson, "Step and flash
        imprint lithography: A new approach to high-resolution patterning," Proc. SPIE
        3676 (1999).

2   *SPL by Electric-Field-Enhanced Oxidation*

2.1   Field-Enhanced Oxidation of Silicon

SPL by electric-field-enhanced oxidation was introduced by Dagata [5] in his pioneering study of patterning hydrogen passivated <111> single crystal silicon with the scanning tunneling microscope (STM). The lithography begins by removing the native oxide and hydrogen passivating the silicon surface in hydroflouric acid (HF). Then the tip of a scanning probe with a voltage bias (typically a few volts) is brought to the vicinity of the surface creating an intense electric field. The magnitude of this electric field can be in excess of 1 V/nm. A schematic diagram of the experimental set up is shown in Fig. 2.1.

Such a large field strength is comparable to that experienced by the electrons orbiting the nucleus. Kreuzer [6] argued that a field of this strength may alter the distribution of electrons between the bonding and antibonding orbitals so that the molecules become more reactive. It is believed that the intense electric field activates the water molecules in air creating negative OH ions and desorbs the hydrogen on the passivated surface [7]. When the probe tip is biased negatively with respect to the sample, the negative oxygen ions are implanted in silicon

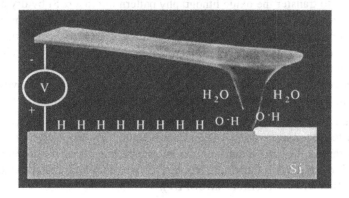

Figure 2.1:   Schematic diagram of SPL by electric-field-enhanced oxidation of silicon.

surface, growing a local oxide. The oxide pattern formed in this manner can serve as an etch mask for transferring the pattern into the silicon substrate.

Since then, Lyding *et al.*[8] has shown hydrogen desorption lines with widths as small as 15 Å using a ultra high vacuum (UHV) STM system. Snow and Campbell [9] demonstrated patterning the (100) silicon surface with the atomic force microscope (AFM) instead of the STM.

2.2 Amorphous Silicon as a Resist Material

A lithographic system capable of patterning only the silicon substrate is of limited value. A more general lithography system based on scanning probes must provide the means to pattern a variety of thin films such as oxides, metals, polymers, as well as semiconductors. Kramer *et al.* [14] showed that since amorphous silicon (α:Si) can be hydrogen passivated in the same manner as single crystal silicon, it is possible to perform electric-field-enhanced oxidation lithography of α:Si with the AFM. Amorphous silicon has many of the characteristics of a general resist material, since it can be deposited as a thin film at moderate temperatures by a variety of means (sputtering, plasma enhanced chemical vapor deposition, and low pressure chemical vapor deposition) on a variety of substrates.

A schematic of SPL using α:Si as a resist is shown in Fig. 2.2. Nominally, we have used 100-nm-thick α:Si deposited by low pressure chemical deposition (LPCVD) as the resist material. The α:Si deposition was performed at 400 mTorr, 560 °C with a 1.24:1 SiH_4: H_2 gas ratio on top of preexisting films such as silicon dioxide or silicon nitride. Alternatively, the α:Si may be deposited by sputtering.

Electric field enhanced oxidation is performed on the α:Si layer by applying a voltage between the AFM probe and the α:Si film. The resulting local oxide is used as the mask to transfer the oxide lithography pattern into the α:Si by dry etching. The etching is nominally done in a 1:1 SF_6:Freon 115 plasma at 150 mTorr. An example of patterned and etched α:Si is shown in Fig. 2.3. Once the pattern is transferred into the α:Si, the underlying film can be patterned with an etch that is selective to the α:Si. Therefore, the maximum etch depth achievable by this technique is governed by the thickness of the AFM-induced oxide, the selectivity of the α:Si etch to oxide, and the selectivity of the subsequent desired film etch to α:Si.

The method of using α:Si as the active patterning layer provides us with the flexibility to patterning a variety of thin films beyond the silicon substrate, with the ability to transfer the pattern much deeper into such films. Using this technique, we proceeded to construct a functional electron device: an n-type metal-oxide-semiconductor field-effect transistor (nMOSFET).

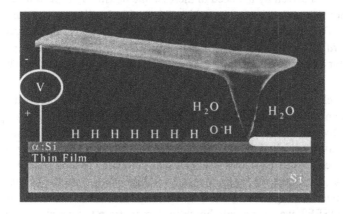

Figure 2.2: Schematic of SPL by electric field enhanced oxidation using amorphous silicon as a resist material.

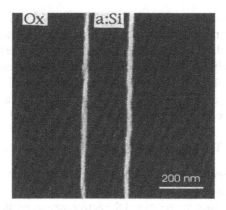

Figure 2.3: Scanning electron microscope (SEM) image of α:Si patterned by SPL with the electric-field-enhanced oxidation. The figure shows a patterned line after etching. The line was patterned at 12 V with 0.5 μm/s scan speed. The line width is 0.2 μm.

2.3 Fabrication of a 100 nm nMOSFET

The quest to scale MOSFETs to smaller dimensions in the semiconductor industry today is driven by the need to increase the density and speed of the digital ICs. The increase in density can be achieved by using smaller gate lengths (L) and widths (W). To increase the speed of digital IC's the saturation drain current in the MOSFET (Id_{sat}) must be increased. The ability to carry more current per width allows faster charging and discharging of capacitances which makes switching faster [15].

In long channel devices (with $L > 2$ μm), Id_{sat} is inversely proportional to L and the thickness of the gate oxide (T_{ox}). Therefore, the motivation for decreasing L and T_{ox} is obvious. However, as gate lengths shrink (< 2 μm), MOSFETs begin to exhibit phenomena called "short channel effects." In short channel devices, Id_{sat} does not increase as much with decreasing L as compared to long channel devices. Nevertheless, the quest for higher density and higher Id_{sat} are the driving forces behind the need for higher resolution lithography. Thus, the fabrication of a short channel MOSFET was a logical choice to demonstrate the capabilities of SPL, including its resolution, its alignment capability, and its compatibility with standard semiconductor device processing.

2.3.1 nMOSFET Design

A two dimensional process simulator, TSUPREM-4,[a] was used to simulate the device and determine most of the processing parameters. The transistor fabrication process uses five levels of lithography, four of which were performed by standard photolithography. Only the patterning of the critical dimension, which is the gate, was performed with SPL. Such an approach of patterning only critical dimension by a different lithography system is often referred to as a "mix and match" approach. A schematic of the process flow is shown in Fig. 2.4.

2.3.2 nMOSFET Process Flow

The nMOSFET process started with a p-type, boron doped, (100) silicon wafer with 10-20 Ω-cm resistivity. For the isolation of the device, 500 nm of field oxide was thermally grown. Next we patterned the active area using standard photolithography (mask level 1) and etched the field oxide in isotropic 6:1 buffered oxide etch (BOE). The inherent undercutting of the isotropic etch ensured that the transition at the edge of the window was not abrupt. A smoothly tapered transition at the edge is important for the gate patterning step by SPL.

---

a. TSUPREM-4 version 5.2.2, Technology Modeling Associates, Inc. Palo Alto, CA.

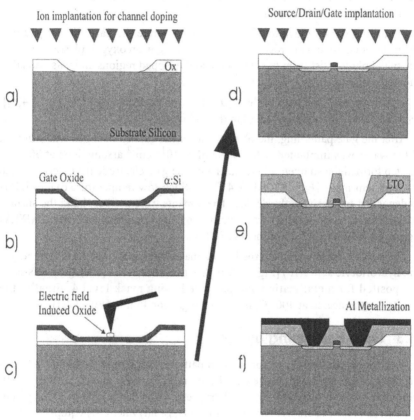

Figure 2.4: Schematic of the process flow for the 100 nm nMOSFET.
(a) Active area is defined and threshold adjust ion implantation is performed.
(b) Gate oxide (8.7 nm) is thermally grown and α:Si is deposited by LPCVD.
(c) Gate is patterned by SPL. (d) Using the oxide from SPL as the mask, α:Si is
etched. The self-aligned ion implantation for source/drain/gate doping is
performed. (e) LTO passivation layer (450 nm) is deposited by LPCVD, and
dopant activation by annealing is performed. Contact openings to source/drain/
gate are defined. (f) Al:Si (1 μm thick) is deposited and patterned before the
forming gas anneal.

The ion implantation step was performed to adjust the threshold voltage and to prevent punch-through between the source and the drain. A 100 keV BF_2 implant was used with a dose of 5×10^{12} cm^{-2} through a 3-nm-thick thermal oxide [Fig. 2.4(a)]. Then a 8.7-nm-thick gate oxide was grown at 850 °C in a dry oxidation furnace. This was followed by a 100-nm-thick α:Si deposition in a LPCVD furnace at 400 mTorr and 560 °C [Fig. 2.4(b)].

Photolithography mask level 2 was used to define the gate contact pad in a 300-nm-thick positive photoresist film. Descumming in an oxygen plasma removed the residual photoresist from the exposed and developed regions and was crucial in achieving a high-quality SPL pattern.

In the next step, the gate of the MOSFET was patterned by SPL [Fig. 2.4(c)]. This procedure is discussed in detail in the following section.

After the gate patterning, the photoresist for the gate contact pad was stripped, and the wafer was implanted with a dose of 1×10^{15} cm^{-2} arsenic ions at 20 keV. This step formed the source/drain regions with the gate electrode itself acting as the self-alignment mask [Fig. 2.4(d)]. A 450-nm-thick low temperature oxide (LTO) was deposited to passivate the device. The dopants were activated and the surface damage from the implantation was repaired in a 10 s rapid thermal anneal (RTA) followed by a 40 min furnace anneal at 800 °C.

The contact holes were patterned using mask level 3, and the LTO was etched using hydroflouric acid (HF) [Fig. 2.4(e)]. A 1-μm-thick Al:Si (1%) alloy was sputter deposited for metallization and patterned using mask level 4. Finally, the contacts were annealed at 400 °C in a forming gas for 45 min [Fig. 2.4(f)].

2.3.3 Gate Patterning by SPL

After the photoresist gate contact pad patterning and descumming, hydrogen passivation was performed on the α:Si by submersing the wafer in a 5:1 H_2O:HF solution for 4 min followed by rinse and dry steps. Then the sample was loaded into a commercial AFM (Park Scientific Instruments, Sunnyvale, CA) equipped with a nitride cantilever coated with 30 nm of titanium. First, the AFM was used in imaging mode with the no voltage bias to register the transistor structure and to determine the location of the gate [Fig. 2.5(a)].

A single pass was made across the active area with a voltage bias of 12 V at a scan speed of 0.55 μm/s. Under these conditions, a 0.21-μm-wide oxide pattern was formed on the α:Si by electric-field-enhanced oxidation. The height of the oxide was around 3.3 nm. In the transition region between the field and the active region, there was approximately 500 nm of topography created by the field oxide. Also, a 300 nm step existed at the edge of the photoresist gate contact pad. In these two regions, two passes were made with the AFM probe at 25 V and 0.55 μm/s to ensure continuity of the gate. We imaged the oxide pattern immediately after writing to verify that the oxide was intact and the coverage was complete. In principle, it is possible to use this mode of imaging to detect and repair defective lithography.

The oxide pattern was transferred into the α:Si using the plasma etch conditions described in Section 2.3. To detect the end-point of the silicon etch accurately, a laser interferometer was used. The plasma was terminated as soon as the α:Si was completely etched. During this etch, the α:Si gate was protected by the oxide pattern and the gate contact pad was masked by the photoresist pattern. The resulting structure is shown in Fig. 2.6. An optical micrograph of the completed nMOSFET is shown in Fig. 2.7. The width of the transistor was 30 μm with a source/drain contact size of 20 μm.

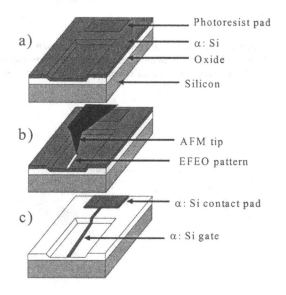

a)
- Photoresist pad
- α: Si
- Oxide
- Silicon

b)
- AFM tip
- EFEO pattern

c)
- α: Si contact pad
- α: Si gate

Figure 2.5: Schematic of nMOSFET gate patterning step by SPL.

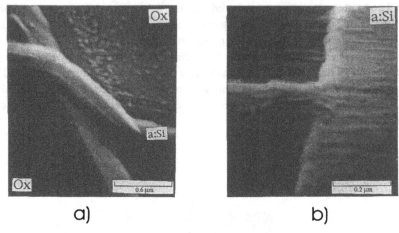

Figure 2.6: SEM micrographs of the patterned amorphous silicon gate.
(a) SEM micrograph of the α:Si gate passing over the transition region between the field and the active areas. (b) SEM micrograph of the gate connection to the gate contact pad.

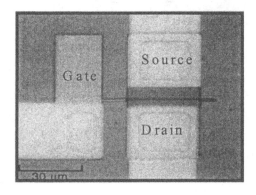

Figure 2.7: Optical micrograph of the completed nMOSFET.

2.3.4 Device Characteristics

Electrical measurements of the nMOSFET were performed on a probe station connected to a HP4145B semiconductor parameter analyzer. The results are shown in Fig. 2.8. For this device, the patterned gate length was 210 nm. We measured the lateral diffusion to be 100 nm through the use of technique described by Chern [16], which was confirmed by the process simulator. The lateral diffusion was caused by the lateral straggle during the self-aligned source/drain implantation as well as the subsequent thermal cycles. Thus our devices had an effective

In Fig. 2.8(a), we plot the drain current (I_d) versus drain-to-source voltage (V_{ds}) with the gate voltage (V_g) varying from 0.5 to 2.5 V. The gate width in this device was 30 μm. We did not measure the switching speed of the devices since the device design was not optimized to reduce the parasitic capacitance of the contact pads and the series resistance of the gate electrode. However, our devices showed

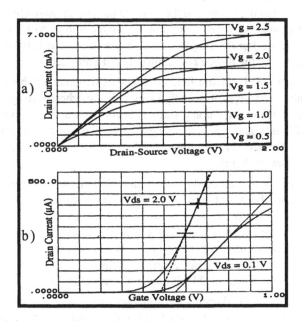

Figure 2.8: Electrical characteristics of the nMOSFET.
(a) Drain current (I_d) versus drain-to-source voltage (V_{ds}) with the gate voltage (V_g) varying from 0.5 to 2.5 V. The width of the gate is 30 μm. The measured saturated transconductance is 279 mS/mm. (b) I_d vs. V_g for V_{ds}=0.1 and 2.0 volts to measure the threshold voltage (V_t). The V_t was determined to be 0.55 and 0.48 V, respectively.

reasonable transconductances compared to short-channel devices fabricated with electron beam lithography. The measured saturated transconductance was 279 mS/mm and the linear transconductance was 145 mS/mm. channel length of 110 nm. All of the measurements were taken at room temperature.

In Fig. 2.8(b), we plot I_d vs. V_g for V_{ds}=0.1 and 2.0 volts to measure the threshold voltage (V_t). The V_t was determined to be 0.55 and 0.48 V, respectively. These measurements are in agreement with the predictions from TSUPREM-4 and PISCES simulators. The subthreshold slopes at V_{ds}=0.1 and 2.0 V are 92 and 110 mV/decade, respectively. The off currents of the device are 16 pA/μm at V_{ds}=0.1 V and 16 nA/μm at V_{ds}=2.0 V.

2.4 Results and Discussion

SPL by electric-field-enhanced oxidation described in this chapter is a powerful lithography technique because of its fine resolution, its capability of *in situ* alignment in the imaging mode, and the resistant oxide etch mask created. We demonstrated the compatibility of this lithography technique with semiconductor integrated circuit (IC) processing by fabricating a functional 110 nm nMOSFET. The characteristics of the device are summarized in Table 2.1.

We believe that SPL electric-field-enhanced oxidation is suitable for patterning a limited number of devices, especially in applications where throughput is not the important factor. However, for applications beyond research and exploration, the viability of SPL as a patterning technology for the semiconductor technology will depend on four factors: (1) resolution, (2) alignment accuracy, (3) reliability, and (4) throughput (wafers/hour).

Considering these factors, the oxidation technique suffers from some major limitations. Firstly, because the oxide formation mechanism is reaction limited at the surface, the writing speed is slow. Snow used a short voltage pulse to oxidize silicon, and from the length of the pulse he inferred a maximum writing speed of 1 mm/s [17]. However, from our experience we could not perform reliable direct patterning at speeds in excess of about 10 μm/s. At these slow scan speeds, patterning of finite chip areas with reasonable throughput is difficult [17].

Secondly, experimental results show that tip wear can be a significant problem. In most of our experiments, we worked with either single crystal silicon or metal-coated tips patterning α:Si or silicon substrates. The tip wear arising from patterning hard surfaces is amplified by the chemical interaction between the tip and the sample as well as the increased force due to the applied voltage. Minne [18] showed that a new silicon tip with a radius of curvature of about 30 nm, will be worn to about 1 μm at its apex after prolonged exposure with a moderate voltage bias of 5 V, at which point lithography is no longer possible. Since the quality of lithography is closely related to the electric field concentration at the tip, which is in turn governed by the tip shape, tip wear poses a serious problem for lithographic reliability.

To overcome the problems of reliability and throughput, while retaining the advantages of fine resolution and accurate alignment capability, we present a different method of SPL in the next chapter. There we show that SPL can be used to pattern electron-sensitive resists using electrons emitted from the probe tip.

Parameter	Value
Physical Characteristics	
Patterned gate length	0.21 μm
Effective gate length	0.11 μm
Width	30 μm
Lithography speed	0.55 μm
Electrical Characteristics	
Transconductance	
Saturated	279 mS/mm
Linear	145 mS/mm
Threshold Voltage	
Vds = 0.1 V	0.55 V
Vds = 2.0 V	0.48 V
Subthreshold Slope	
Vds = 0.1 V	92 mV/decade
Vds = 2.0 V	110 mV/decade
Off Current	
Vds = 0.1 V	16 pA/μm
Vds = 2.0 V	16 nA/μm

Table 2.1: Device characteristics of an nMOSFET patterned with SPL by electric-field-enhanced oxidation.

2.5 References

[1]   J.S. Kilby, "Invention of the integrated circuit," IEEE Transactions on Electron Device, 648 (1976).

[2]   R. Mohondro, "Advanced lithography: A review," Future Fab International **1**, 121 (1996).

[3]   The information is available from The National Technology Roadmap for Semiconductors 1997 Edition, SEMATECH, Austin TX, 78758 (1997).

[4]   R. Stulen, "Technical challenges in extreme ultraviolet lithography," Proceedings of the Sixth International Symposium on Ultralarge Scale Integration Science and Technology, 515 (1997).

[5]   J. A. Dagata, J. Schneir, H. H. Harary, C. J. Evans, M. T. Postek, and J. Bennett, "Modification of hydrogen-passivated silicon by a scanning tunneling microscope operating in air," Appl. Phys. Lett. **56**, 2001 (1990).

[6]   H .J. Kreuzer, "Physics and chemistry in high electric fields," Atomic and Nanometer-Scale Modification of Materials: fundamentals and Applications, Ed: Phaedon Avouris, NATO ASI Series, Series E, Applied Science **239**, 75-86 (Boston: Kluwer Academic Pub., 1993)

[7]   R. S. Becker, G. S. Higashi, Y. J. Chabal, and A. J. Becker, "Atomic scale conversion of clean Si(111):H-1x1 to Si(111)-2x1 by electron-stimulated desorption," Phys. Rev. Lett. **65**, 1917 (1990).

[8]   J. W. Lyding, "Nanoscale patterning and oxidation of H-passivated Si(100)-2x1 surfaces with an ultrahigh vacuum scanning tunneling microscope," Appl. Phys. Lett. **64**, 2010 (1994).

[9]   E. S. Snow, and P. M. Campbell, "Fabrication of nanometer-scale side-gated silicon field effect transistors with an atomic force microscope," Appl. Phys. Lett. **66**, 1388 (1995).

[10]   H. Sugimura, T. Uchida, N. Kitamura, and H. Masuhara, "Tip-induced anodization of titanium surfaces by scanning tunneling microscopy: A humidity effect on nanolithography," Appl. Phys. Lett. **63**, 1288-1290 (1993).

[11]   T. Thundat, L .A. Nagahara, P. I. Oden, S. M. Lindsay, M. A. George, and W. S. Glaunsinger, "Modification of tantalum surfaces by scanning tunneling microscopy in an electrochemical cell," J. Vac. Sci. Technol. A **8**, 3537 (1990).

[12]   H. J. Song, M. J. Rack, K. Abugharbieh, S. Y. Lee, V. Khan, D. K. Ferry, and D. R. Allee, "25 nm chromium oxide lines by scanning tunneling lithography in air," J. Vac. Sci. Technol. B **12**, 3720 (1994).

References

[13]   E. S. Snow, D. Park, and P.M. Campbell, "Single-atom point contact devices fabricated with an atomic force microscope," Appl. Phys. Lett. **69**, 269 (1996).

[14]   N. Kramer, H. Birk, J. Jorritsma, and C. Schonenberger, "Fabrication of metallic nanowires with a scanning tunneling microscope," Appl. Phys. Lett. **66**, 1325 (1995).

[15]   Y. Tsividis, *Operation and Modeling of the MOS transistor* (New York: McGraw-Hill, Inc., 1987), p. 168.

[16]   J. G. Chern, P. Chang, R. F. Motta, and N. Godinho, "A new method to determine MOSFET channel length," IEEE Electron Device Lett. **EDL-1**, 170 (1980).

[17]   E. S. Snow and P. M. Campbell, "Fabrication of Si nanostructures with an atomic force microscope," Appl.Phys. Lett. **64**, 1932 (1994).

[18]   S. C. Minne, "Increasing the throughput of atomic force microscopy," Ph.D. Thesis, Stanford University (1996).

3   *Resist Exposure Using Field-Emitted Electrons*

Early scanning probe lithography (SPL) studies were limited to demonstrations of the technique's fine resolution. A few groups fabricated devices using SPL [1][2][3], but such work was directed toward creating a single working device suitable for research or exploration. Methods used by these groups suffer from speed constraints and poor repeatability, thus it is unlikely they can be easily extended to large-scale fabrication applications. We sought to develop a method of SPL suited to semiconductor lithography, where accuracy, reliability, and throughput are essential.

This chapter presents our methods of reliable high-resolution patterning using electrons emitted from a scanning probe. In Section 3.1 we describe the main elements of an SPL system, namely the electron emission, resists, substrates, tips, writing strategies, and pattern transfer methods. We provide a summary of the characteristics and limitations of resist exposure using "traditional" SPL techniques. We show that the electron exposure dose is the critical parameter for patterning organic resists and present two SPL techniques—a contact mode system (Section 3.2) and a noncontact mode system (Section 3.3)—in which we incorporate real-time control of the exposure dose. We demonstrate high-resolution patterning and good pattern transfer fidelity using these systems. Section 3.4 presents results of modeling the electron field emission from a biased probe tip.

3.1 Field-Emitted Electron Exposure

The main elements of an SPL system are illustrated in Fig. 3.1(a). A sharp tip is positioned in close proximity to a resist-coated sample. The tip is biased negatively with respect to the sample. Electrons emitted from the tip transfer energy to the resist film, inducing chemical changes in the resist. In this section we describe these SPL components in more detail.

3.1.1 Electron Field Emission

A voltage bias between a sharp tip and a flat electrode generates an intense electric field concentrated at the tip. Figure 3.1(b) shows the electric field distribution near a biased probe tip. Here the tip has a 10 nm radius of curvature and a 20° cone angle. For a tip-to-sample spacing of 50 nm and an applied bias of 50 V, the maximum electric field is located at the tip apex and has a magnitude of about 2.2 V/nm. This is comparable to the field strength experienced by electrons orbiting the nucleus of an atom [4].

This high electric field can deform the potential barrier at the tip surface. When the barrier becomes sufficiently narrow, there is a finite probability that electrons in the emitter's Fermi sea can quantum-mechanically tunnel out of the tip (Fig. 3.2) [5]. This process is known as electron field emission and was first explained by R. H. Fowler and N. Nordheim in 1928 [6]. The field-emitted current depends critically on the emitter work function and the field strength at the tip, as given by by the Fowler-Nordheim equation [7]:

$$J = 1.5 \times 10^{-6} \frac{E^2}{\phi} \exp\left(\frac{9.87}{\phi^{\frac{1}{2}}}\right) \exp\left(\frac{-6.53 \times 10^7 \phi^{\frac{3}{2}}}{E}\right) \qquad . \quad (3.1)$$

J is the emitted current density (in A/cm^2), E is the magnitude of the electric field normal to the tip surface (in V/cm), and ϕ is the emitter work function (in eV).

Field emission in air is rather unstable as adsorbates on the surface of the tip cause local fluctuations in the work function [8]. Also, ions can sputter the surface and change the emission efficiency. As a consequence, field-emission cathodes are typically operated under high vacuum to ensure stable emission currents [9]. Field-emission sources are often used for scanning electron microscopy. Recently arrays of field-emitting tips have been used in high-resolution, high-brightness flat panel displays [10]. We use the field-emitted electrons to expose electron-sensitive resists.

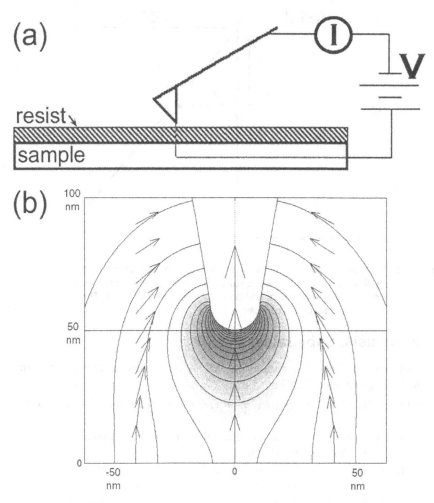

Figure 3.1: (a) Schematic diagram of a cantilever with integrated tip used for electron emission and resist exposure. (b) Electric field distribution near a biased probe tip. The tip has a 10 nm radius of curvature and a 20° cone angle. The tip and sample are separated by 50 nm of air. The tip is biased at 0 V and the sample at +50 V. The shading corresponds to the electric field magnitude and the lines are electric field contours. The arrows show the direction of the electric field vector.

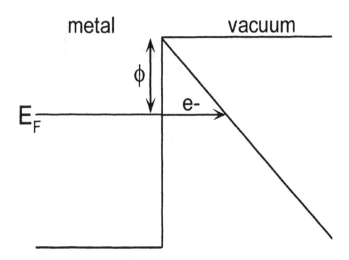

Figure 3.2: Potential energy diagram for electrons at a metal surface in the presence of an applied field. When the field is sufficiently strong, electrons can leak through the deformed potential barrier and be emitted from the conductor. This process is known as electron field emission.

3.1.2 Resists, Substrates, and Tips

There are four important parameters that we must consider in evaluating the potential of a given resist for SPL:

(1) resist threshold energy (minimum electron energy required to expose the resist);

(2) resist sensitivity (minimum electron dose necessary to expose the resist at energies above the threshold);

(3) resist resolution limit (minimum feature size that can be printed on a given resist, generally limited by the molecular size of the resist); and

(4) resist etch selectivity (resistance to dry etches, which sets the minimum thickness of the resist that can be used for pattern transfer into various films).

Organic polymer resists provide an attractive option for SPL because many have a low threshold energy, high sensitivity, sub-100-nm resolution, and good dry etch resistance.

We employed commercially-available organic polymer resists used tradition-ally for electron beam lithography (EBL). We spin-coated the resists onto 4-inch-diameter silicon wafers. In order to achieve thin resist films (35–100 nm thick), we adjusted the resist viscosity (by diluting the resists in an appropriate solvent) and spun the solution at high speeds. For a spun resist thickness down to 35 nm we saw no evidence of pinhole formation in the resist film, although a more thorough investigation of defects in thin resists is certainly needed.

We used SPL to pattern Microposit SAL601, a negative-tone electron-beam resist from Shipley Company. SAL601 is a high-contrast novolac-based organic polymer resist. The high resolution of SAL601 (< 50 nm) was previously demonstrated with x-ray [11], electron beam [12], and STM [13] lithography. SAL601 is a chemically-amplified resist (CAR), which makes it highly sensitive to electron exposure. Generally, exposure of a CAR by an incident energy source (such as electrons, ions, or DUV light) generates a photoacid in the resist. This acid causes further chemical reactions during a high-temperature activation step (post-exposure bake, PEB) leading to cross linking of the matrix polymer. The exposed SAL601 regions remain after development, thus the resist patterns are straightforward to characterize.

#	Step	Equipment	Temp	Time
1	Oxide etch[a]	50:1 HF	RT[b]	35 s
2	Singe	Convection Oven	150 °C	30 min
3	Prime	HMDS Vapor	125 °C	30 s
4	Resist spin	Thinned SAL601	RT	30 s
5	Soft bake	Hot plate	85 °C	1 min
6	Pre-bake	Convection Oven	90 °C	30 min
7	Exposure	SPL	RT	≤ 2 h
8	PEB	Hot plate	115 °C	1 min
9	Development	MF-322	RT	10 min
10	Post-bake	Hot plate	115 °C	1 min

Table 3.1:   Sample preparation steps for SPL patterning of SAL601 resist.

a. For silicon or polysilicon samples only.
b. RT=room temperature.

We patterned SAL601 on substrates of silicon, polysilicon, and metal. The sample preparation steps are summarized in Table 3.1. For the silicon and polysilicon samples, the native oxide was etched off the substrate in 50:1 hydrofluoric acid (HF) prior to resist coating. The wafers were then singed for 30 min in a 150 °C convection oven and vapor primed with hexamethyldisilazane (HMDS) adhesion promoter. SAL601 diluted in Microposit Thinner Type A was spin coated on the wafers.[a] Then the wafers were baked (1 min, 85 °C hot plate followed by 30 min, 90 °C convection oven). We measured the resulting resist thickness using the Nano-Spec/AFT Model 210XP spectrophotometer. All above steps were performed in quick succession in a Class 100 clean room environment.[b]

Because of this resist's sensitivity to contamination and to delay times, wafers were prepared with SAL601 just prior to exposure. Immediately following exposure the wafers were given the PEB (1 min, 115 °C hot plate) and then submersed in MF-322 developer for 10 min, rinsed in de-ionized water, and blown dry with nitrogen. The final step was a post-bake (1 min, 115 °C hot plate) to harden the resist. When the delay time between the pre-bake and development was greater than about two hours we noticed delamination of the resist lines. After development, the substrate surface (silicon or polysilicon) should be hydrophobic. For long delay times the surface was often hydrophyllic, indicating that oxygen had diffused through the resist and grown a native oxide on the substrate. We attribute the resist delamination to this oxide growth, since SAL601 does not adhere well to oxide.

We also used SPL to pattern polymethylmethacrylate (PMMA), a well-known positive-tone organic polymer resist. PMMA is an ultra-high-resolution resist (patterns as small as 10 nm have previously been created in PMMA by EBL [14]). Prior to resist coating, the native oxide was removed from the silicon or polysilicon samples and the wafers were singed at 150 °C (Table 3.2). A solution of 1.25% PMMA[c] by weight in chlorobenzene (C_6H_5Cl) was spin coated on the wafers. The wafers were baked for 2 h at 170 °C in a convection oven. After SPL exposure, the PMMA was developed for 1 min in a high contrast developer solution whose components are listed in Table 3.3 [15][16]. This developer is designed to enhance the resist contrast and provide appropriate resist sidewall profiles for lift-off. After development, the samples were rinsed in isopropanol and blown dry with nitrogen.

A voltage between the tip and the film beneath the resist enables electron emission from the tip. SPL therefore requires a conducting (or semiconducting) underlayer. The film does not have to be highly conductive because it is in series with the high impedance resist, but it should not be an insulator. Substrates such as doped silicon, doped polysilicon, and metal can be directly patterned. A bilayer or trilayer resist system can be used to pattern insulating layers like interlevel dielectrics [17]. For example, a conducting pattern transfer layer like germanium (Ge) or metal could be deposited on top of the insulating film to be patterned. The resist would be coated on top of the transfer layer and the sample bias applied to the

---

a. For example, a 1:3 ratio of SAL601 resist to Thinner A spun at 7 krpm yielded a 65-nm-thick resist film (after baking).
b. The Stanford Nanofabrication Facility (SNF) at Stanford University.
c. PMMA molecular weight 950 000 amu.

---

transfer layer. Developed resist features would be etched first into the Ge or metal, and next into the underlying insulating film using the transfer layer pattern as the etch mask.

We used high-aspect-ratio doped-silicon tips as the electron emitters.[a] We coated some silicon tips with a thin film of metal (such as titanium, tantalum, or molybdenum), carbon, or diamond, which covered well the full length of the tip. The radius of the silicon tip was about 10 nm. With the metal coating, the tip radius was approximately 30-40 nm. See Chapter 9 for more discussion of probe tips.

#	Step	Equipment	Temp	Time
1	Oxide etch[a]	50:1 HF	RT	35 s
2	Singe	Convection Oven	150 °C	30 min
3	Resist spin	Thinned PMMA	RT	30 s
4	Pre-bake	Convection Oven	170 °C	2 h
5	Exposure	SPL	RT	--
6	Development	Liquid Developer[b]	RT	1 min
7	Rinse	Isopropanol	RT	1 min

Table 3.2:   Sample preparation steps for SPL patterning of PMMA resist.

a. For silicon or polysilicon samples only.
b. See Table 3.3.

Component	% Volume
Isopropanol	37.5
Methanol	31.8
2-Ethoxyethanol	13.6
Methyl Iso-Butyl Ketone	12.5
Ethyl Alcohol	3.3
Methyl Ethyl Ketone	1.2

Table 3.3:   Components of the PMMA high contrast developer.

a. Ultralevers™ from Park Scientific Instruments (Sunnyvale, CA) or custom fabricated probe tips that were etched in silicon and oxidation sharpened.

3.1.3 Writing Strategies

There are two common writing strategies adopted for direct-write lithography, both of which can be used for SPL. The first is *raster scanning* in which the probe is scanned over every pixel on the sample but the exposure (voltage or current, for example) is enabled only over pixels to be written [Fig. 3.3(a)]. Alternatively, *vector scanning* may be adopted, where the probe is driven only along the path to be written [Fig. 3.3(b)].

With one tip, vector scanning is often simpler and more efficient since no time is "wasted" scanning over regions where no patterns will be written. This is most often used for SPL where the voltage or current is enabled as software controls the path and speed of the tip. Raster scanning is a more general approach and may be necessary if multiple probes are scanned simultaneously for parallel lithography. An array of probes could be raster scanned over the sample as a unit and the exposure enabled independently for each probe to write arbitrary patterns.

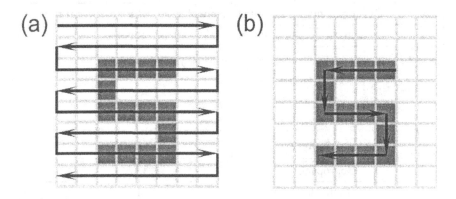

Figure 3.3: Depiction of two different scanning strategies for direct write lithography. (a) Raster scanning. (b) Vector scanning.

3.1.4 Pattern Transfer Techniques

We transferred the developed resist patterns into the silicon substrate through direct etching or lift-off. The SAL601 exposure, develop, and etch steps are illustrated in Fig. 3.4(a). The patterned resist was hard-baked at 110 °C for 30 min prior to etching. We performed direct etching in a Drytek 100 plasma etcher or a LAM Research Systems TCP9400 reactive ion etcher (RIE). In the Drytek, SF_6 + F115 chemistry was used to achieve reasonably anisotropic pattern transfer with a selectivity of about 3:1 of silicon to SAL601. Anisotropy is achieved in this etcher through polymer redeposition on the etched feature sidewalls. The newer LAM etcher generates a high density plasma using $HBr + O_2$ gases and achieves highly anisotropic etch profiles. This etch has a selectivity better than 5:1 of silicon to SAL601. PMMA has poor dry etch resistance (~1:1 selectivity to silicon in the LAM $HBr + O_2$ etch discussed above). Therefore it is useful to employ a lift-off technique to transfer a thin PMMA pattern deep into the underlying film [Fig. 3.4(b)]. Lift-off is also useful for reversing the tone of the pattern. We implemented this transfer technique by evaporating 10 nm of chrome onto the wafers after PMMA development. The chrome adhered to the silicon in the regions where the resist was exposed and developed away; it coated the top surface of the resist wherever PMMA remained. The chrome was lifted off in acetone (which dissolved the remaining resist) using ultrasonic agitation, and then the wafers were rinsed in methanol.

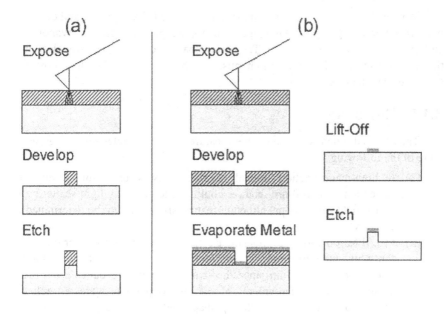

Figure 3.4: Schematic diagrams of SPL resist exposure, development, and pattern transfer. (a) Negative tone resist patterning and pattern transfer through direct etching. (b) Positive tone resist patterning and pattern transfer through metal evaporation, lift-off, and etching.

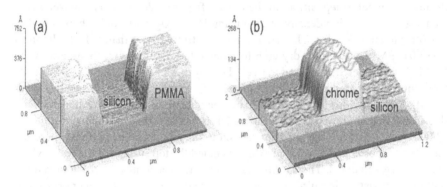

Figure 3.5: Transfer of PMMA patterns through chrome evaporation and lift-off. (a) AFM image of developed resist pattern on silicon. (b) AFM image after chrome evaporation and lift-off.

An AFM image of the sample topography after PMMA exposure and development is shown in Fig. 3.5(a). An image of the topography after chrome evaporation and lift-off is shown in Fig. 3.5(b). The chrome pattern can be transferred deep into the silicon using an RIE with NF_3 chemistry.[a] This etch is highly anisotropic and selective against chrome.

3.1.5 Metrology

The SPL-patterned resist and etched features were characterized using one or more of the following:

(1) A high-magnification optical microscope was used for preliminary inspection. Sub-100-nm features could be identified by light scattering in a dark-field microscope, although feature size could not be determined in this way.

(2) An AFM operating in noncontact mode was used to determine line continuity and uniformity. Contact-mode operation was not acceptable for resist pattern inspection since the scanning tended to damage the resist features. AFM measurements of the etched patterns were not reliable because of convolution of the tip shape and the high-aspect-ratio etched features.

(3) The Hitachi S-800 scanning electron microscope (SEM) operating at 25 keV was used for inspection and line width measurements. Images and measurements were made from top-down, tilted, and cross-section images.

---

a. AMT 8100 Hexagonal etcher run with NF_3 flow=20 sccm, pressure=20 mT, self bias=-430 V, and power=1400 W. Silicon etch rate ~ 300 Å/min.

---

(4) The Dektak SXM critical dimension AFM (CD-AFM) was used for measuring line widths, line heights, and sidewall profiles. We operated the CD-AFM in a two-dimensional noncontact scanning mode using a boot-shaped tip [18]. Previous studies showed that CD-AFM measurements correlate well with cross-section SEM data [19]. An advantage of the CD-AFM is that it is capable of making these measurements non-destructively (without cleaving the wafer and preparing a cross section), and measurements and images can be obtained anywhere on the wafer (not just where the cross section was made).

3.1.6 Resist Patterning With the AFM or STM

Organic polymer resists have been patterned previously by both the STM and AFM. For STM lithography [Fig. 3.6(a)], a fixed voltage bias between an etched tungsten wire tip and a sample generates the field emission of electrons from the tip. The spacing between the tip and sample is varied to maintain a constant current through the resist [13][20][21]. The STM has been used to write sub-30-nm features in resist. This system, however, suffers from poor alignment capabilities since imaging may expose the resist [22]. It is also tricky to determine the appropriate voltage bias for STM lithography. If the bias is set too high, the tip may move far from the sample resulting in spreading of the beam and degradation of the resolution. If the bias is too low, the tip may penetrate the resist in an attempt to achieve the setpoint current [23]. These factors make reliable STM lithography of resists problematic.

In AFM lithography [Fig. 3.6(b)], the force between the tip and resist is held constant while a fixed voltage bias is applied to generate the field-emitted current [24]. The distance between the tip and sample is minimized, enabling high-resolution patterning. Yet constant-voltage operation is not ideal since the voltage required for emission may depend on the tip and sample materials, the tip shape, and the resist thickness. Any change in tip shape or resist thickness, for instance, varies the dose of electrons delivered to the resist. This results in nonuniform and unrepeatable patterns. We measured the emission current during constant voltage AFM lithography and found it was extremely unsteady. Line width variations in the patterned features correlated with the emission current fluctuations.

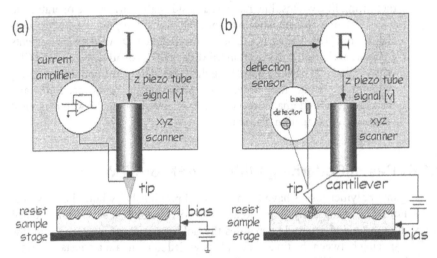

Figure 3.6:   Schematic diagrams of resist exposure using electrons field emitted from a scanning probe in the (a) STM configuration and (b) AFM configuration.

3.2   Current-Controlled Exposures in Contact Mode

We developed a new SPL technique that combines the key features of the AFM and STM for reliable patterning. Our lithography system is capable of patterning sub-100-nm features in organic resist on a variety of substrates, with various tips, and over topography. It requires minimal characterization prior to patterning.

3.2.1   The Hybrid AFM / STM Lithography System

The "hybrid AFM/STM" system incorporates two independent feedback loops, one that controls the tip-sample force and one that controls the field-emission current from the tip (Fig. 3.7). The force feedback loop is precisely that used in an AFM imaging tool. Its role is to keep the tip on the surface of the electron-sensitive resist and to maintain a constant small force between the tip and resist surface (typically ~ 10 nN). In our system, we detected the tip-sample force by monitoring the cantilever deflection with a conventional optical lever beam-bounce system. Other deflection sensors, such as an integrated piezoresistive sensor [25] or an interdigital cantilever [26], might also be used. A voltage signal was sent to the piezotube actuator to move the probe up or down to maintain the setpoint force.

The second feedback loop maintains a constant field-emission current through the resist during exposure (typically in the range of 10 pA to 50 nA) by adjusting the voltage bias between the tip and sample. A custom-built analog feedback circuit

(Fig. 3.8) uses integral gain to minimize the current error signal. The input stage in the feedback circuit compares the measured current to the setpoint current. A high voltage amplifier (Apex MicroTechnology PA85) applies the necessary voltage to the sample (typically 20-100 V) to maintain the setpoint current. We measured the current through the resist with picoAmpere sensitivity using a commercial Ithaco 1211 current preamplifier. The current was sensed at the tip for lowest noise measurements. Feedback bandwidth was limited to roughly 1 kHz for maximum loop gain.

We performed hybrid AFM/STM exposures in air using the Park Scientific Instruments (PSI) AutoProbe M5 operating in contact AFM mode modified with the constant-current feedback circuitry. A piezotube scanner with a reach of 100 μm was used to scan the tip over the sample, and a full wafer stage was used for sample positioning. We mounted the tip on an insulating holder and made electrical connections to the cantilever chip with a thin line of silver paint (this minimized the tip-sample capacitance). This capacitance becomes important when attempting to maintain constant current at high speeds or over topography (see Chapter 6). We applied a positive voltage to the sample with a metal clip and grounded the tip through the current preamplifier. We used PSI Nanolithography software to specify the path and the speed of the tip. Patterns were written at a variety of scan speeds and exposure doses, with different tips and on different resists and substrates.

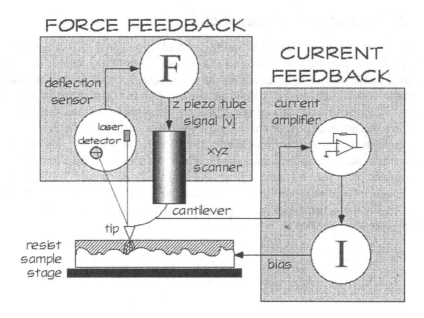

Figure 3.7: Schematic diagram of the hybrid AFM/STM lithography system. One feedback loop maintains a constant force between the tip and sample while the other independent loop keeps a constant field-emission current through the resist.

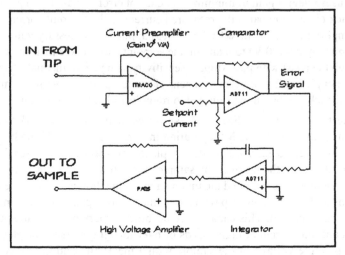

Figure 3.8:   Diagram of the integral feedback circuit used for emission current control. The
input stage converts the measured current to a voltage. This value is compared
with the setpoint level. The high voltage output amplifier adjusts the sample bias
in order to maintain the setpoint exposing current.

3.2.2 Emission Characteristics

Figure 3.9 shows the dependence of the emitted current on the tip-sample voltage bias for electron emission through SAL601 resist films of thickness 35 nm and
65 nm. Data were acquired by measuring the applied bias necessary to achieve each
current level (with active current feedback circuitry) while scanning the tip in contact with the resist surface at a speed of 10 μm/s. For a spun resist thickness of 35
nm, there was no measured current below 17 volts. Once the field at the tip became
high enough for current to flow (above this 17 volt "threshold"), the current
increased dramatically with voltage. The threshold voltage for emission through the
65-nm-thick resist film was nearly 42 volts. From the slope of the I-V curves, we
find that once the current starts to flow the resist film has an effective impedance of
> 10 GΩ. We found that the voltage required to emit a given current also varies with
writing speed (see Section 6.2.3 for details).

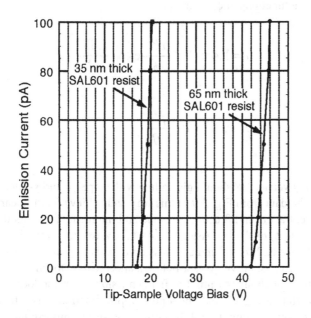

Figure 3.9: Emission current as a function of the tip-sample voltage bias for SAL601 resist thicknesses of 35 nm and 65 nm. The data were acquired using the hybrid AFM/STM system by stepping the setpoint current and monitoring the applied voltage necessary to achieve that setpoint. The Ti-coated tip was in contact with the resist surface.

We expect that the I-V relationship depends also on the tip shape. For example, a more blunt tip would require a higher bias to reach the critical field for electron emission. It is difficult to achieve uniform patterning in a constant voltage SPL mode because of the strong dependence of emitted current on the tip-sample voltage bias, resist thickness, scan speed, and tip shape.

The shape of the curves in Fig. 3.9 is described reasonably well by the Fowler-Nordheim field-emission theory [6]. Following the analysis by Spindt and Brodie [7], we substitute the approximations $J=I/\alpha$ and $E=\beta V$ into Eq. (3.1), where α is the emitting area and β is known as the field enhancement factor. This yields a relationship of the form:

$$I = aV^2 \exp\left(-\frac{b}{V}\right) \qquad , \qquad (3.2)$$

where a and b are functions of ϕ, α, and β:

$$a = 1.4 \times 10^{-6} \frac{\alpha \beta^2}{\phi} \exp\left(\frac{7.832}{\phi^{\frac{1}{2}}}\right) \qquad , \qquad (3.3)$$

$$b = 6.53 \times 10^7 \frac{\phi^{\frac{3}{2}}}{\beta} \qquad . \qquad (3.4)$$

A plot of $ln(I/V^2)$ versus $1/V$ should give a line with slope $-b$ and y-intercept $ln(a)$. We re-plotted the data from Fig. 3.9 in this way and achieve good linear fits (Fig. 3.10). We can extract α and β using the slope and intercept of the fits if we know the work function, ϕ, of the tip. Assuming ϕ=4.3 eV for titanium, we find a field enhancement factor of 2.0×10^6 and an emission diameter of about 7 Å for the thin resist. The thicker resist curve gives a field enhancement factor of 5.4×10^5 and an emission diameter of about 147 nm. For field emission in air, the local work function may vary, thus ϕ=4.3 eV may be a poor approximation. Also, this analysis assumed $J=I/\alpha$, which is only true if the current density is uniform throughout the emission area. Nevertheless, we see that the emission area seems to increase as the tip-sample distance increases (see Section 3.4 for more discussion)

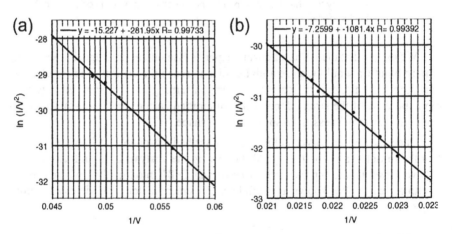

Figure 3.10: Fowler-Nordheim plots of $ln(I/V^2)$ versus $1/V$ for the data shown in Fig. 3.9. The current was field-emitted through (a) 35-nm-thick and (b) 65-nm-thick SAL601 resist.

3.2.3 Latent Images in the Resist

Immediately following SPL exposure, we examined the resist surface topography using the AFM. We disabled the current feedback (simply by setting the current setpoint to zero) and scanned the exposed area. Latent images were evident in both the SAL601 and PMMA resist films. In SAL601, the exposed regions were raised as much as 10 nm over the unexposed regions, where the amount of swelling depended on the exposure conditions. Fig. 3.11(a) shows the SAL601 resist topography after patterning a 61-nm-thick film with an exposure dose of 1000 nC/cm. The exposed areas were raised about 4.5 nm over the unexposed regions. The thickness of the lines was even greater after development, suggesting additional resist swelling during the post exposure bake (PEB) and/or development steps. Table 3.4 lists the resist thickness change from spin-on to development for four different initial SAL601 resist thicknesses. All were exposed with the same dose. The spun resist thickness was measured by the NanoSpec/AFT Model 210XP spectrophotometer and the developed thickness by the CD-AFM. The increased swelling for the thicker resist is likely due to the higher bias necessary to achieve the setpoint current.

Negative resists tend to swell during development, creating characteristic reentrant profiles [27]. Swelling during exposure is not generally seen for electron beam lithography (EBL) or deep ultraviolet (DUV) exposure of CARs. Ocola *et al.* studied the latent image formation for EBL of SAL605, a negative tone DUV CAR manufactured by Shipley with similar chemistry as SAL601 [28]. Their results showed that before the PEB the exposed regions were valleys in the resist, while afterwards the features became ridges. In other words, the release of acid in the resist induced by exposure *reduced* the resist volume and the cross linking induced by the PEB *increased* the resist volume.

We found that SPL exposure can itself cause cross linking of the SAL601 matrix polymer. In fact, no PEB step is necessary to create features with SPL. Therefore it is consistent that we observed raised latent images in the resist immediately after SPL exposure. Our results agree with those reported by Perkins *et al.* [29] for STM exposure of SAL601 and by Shiokawa *et al.* [30] for constant voltage AFM exposures of SAL601.

We also observed latent images in the PMMA positive tone resist film following exposure but before development. Interestingly, the exposed regions were again raised over the unexposed regions despite the fact that PMMA and SAL601 are opposite tone resists. Figure 3.11(b) shows an AFM image and the associated height profile of the PMMA resist surface topography immediately after exposure. The raised regions dissolved away during the subsequent development step.

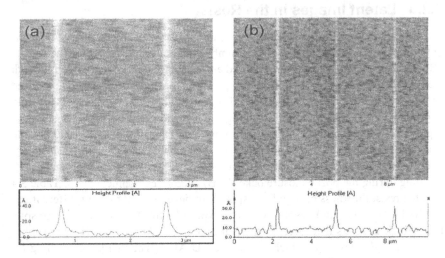

Figure 3.11: AFM micrographs of latent images in resist immediately after SPL exposure. (a) Lines patterned with an exposure dose of 1000 nC/cm in 61-nm-thick SAL601 negative resist. The exposed areas are raised about 4.5 nm above the unexposed regions. (b) Lines patterned in 50-nm-thick PMMA positive resist. The exposed area are raised about 2.5 nm above the unexposed regions.

The AFM's ability to image latent features formed in the resist film during exposure suggests the feasibility of *in situ* monitoring of the lithography process and/or aligning with previously written features. Pattern correction might also be performed before development [30].

$T_{initial}$ (nm)	T_{final} (nm)	Δ_{fi} (nm)	V_t (V)
36.3	42.0	5.7	21
42.1	53.0	10.9	32
53.2	65.0	11.8	35
60.7	74.6	13.9	46

Table 3.4: SAL601 resist swelling for different initial resist thicknesses. Initial resist thickness after spin-on and baking ($T_{initial}$), final resist thickness after exposure and development (T_{final}), and the amount of resist swelling (Δ_{fi}) are shown. Also listed is the field-emission threshold voltage (V_t) dependence on resist thickness.

3.2.4 Resolution and Line Width Control

The current feedback system of the hybrid AFM/STM ensures that a fixed number of electrons is delivered to the resist per unit time. The exposure line dose is the ratio of the exposing current to the writing speed and has units of charge per unit length (typically nC/cm). Figure 3.12 shows gratings with lines on a 500-nm pitch (line spacing) written by SPL in SAL601 and etched into the underlying silicon substrate. Each line was written in a single pass of the tip. The lines in Fig. 3.12(a) were written with an exposure dose of 40 nC/cm and have a width of 52 nm. When the exposure dose was increased by a factor of five to 200 nC/cm, the line width increased to 123 nm [Fig. 3.12(b)]. The features exhibit excellent uniformity, good line width repeatability, and smooth line edges. The developed resist features were slightly re-entrant, with sidewall angles of approximately 93° to 110°, as measured by the CD-AFM.

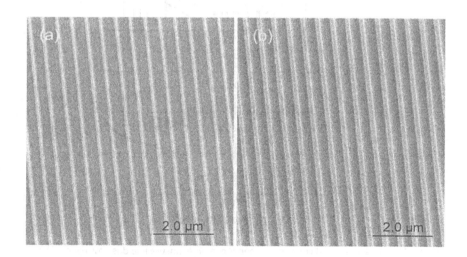

Figure 3.12: SEM micrographs of 500-nm-pitch line gratings patterned by SPL. Patterns were written in SAL601 resist and directly etched into the underlying silicon substrate. (a) SPL exposure line dose of 40 nC/cm yielded line width of 52 nm. (b) SPL exposure line dose of 200 nC/cm yielded line width of 123 nm.

We varied the exposure dose and measured the resulting line width for SPL lithography of SAL601 and PMMA resists on doped silicon substrates (Fig. 3.13). The dose was varied by adjusting either the exposing current, I, or the scan speed, s (where the ratio of I/s is the relevant line dose). The same line width versus dose relationship shown in Fig. 3.13 for SAL601 exists for both (1) a constant emission current of 50 pA and a variation of writing speed from 1–50 μm/s, and (2) a constant writing speed of 10 µm/s and a variation of emission current from 10–500 pA. In both cases the exposure dose was varied from 10-500 nC/cm. From this data we conclude that the critical parameter for exposing an organic polymer with scanning probes is the dose of electrons delivered to the resist. The width of the exposed region depends critically on this line dose.

PMMA requires a threshold dose 20 times higher than that needed by SAL601, indicating that it is less sensitive to the low-energy electrons.[a] At the appropriate doses we printed features with widths below 50 nm in both resists. We achieved smaller lines over a wider dose range using PMMA.

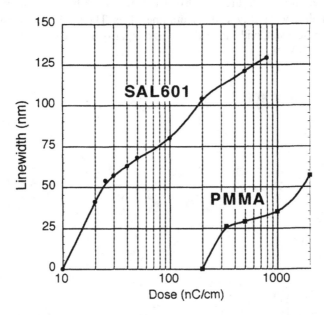

Figure 3.13:   SPL patterned line width dependence on exposure line dose for hybrid AFM/ STM patterning of 65-nm-thick SAL601 resist and 50-nm-thick PMMA resist on doped silicon substrates.

Line width measurements of patterns written with different tips and on different samples fell within 10% of the values on the curves in Fig. 3.13. We did not observe any change in the dimension of patterns written after extended tip use. Terris *et al.* used a silicon tip in contact with a polymer-coated spinning disk and read data bits continuously for over 145 h, equivalent to a tip travel distance of 16 km,

---

a. PMMA is also less sensitive to high-energy electrons from an EBL system than is SAL601. See Section 3.2 for details.

without any significant change in signal amplitude [32]. This result and our experience suggest that the surface of an organic resist is sufficiently soft and pliable to prevent tip damage.

The real-time exposure dose control of hybrid AFM/STM lithography also enables uniform patterning over sample topography where the resist thickness changes as a function of position. This capability is discussed in more detail in Section 5.2.2.

3.2.5 Pattern Transfer

The high selectivity and anisotropy of the reactive ion etches (RIE) described in Section 3.1.4 allow us to transfer the narrow resist patterns deep into the underlying film or substrate to create high-aspect-ratio features. Figure 3.14(a) shows an optical micrograph of developed SAL601 resist lines on polysilicon. These patterns were transferred into the 100-nm-thick polysilicon film using the Drytek SF_6 / F115 anisotropic dry etch and then the resist was stripped. Fig. 3.14(b) shows the resulting etched lines. Fig. 3.14(c), taken at high magnification, illustrates the excellent line uniformity and continuity of the features. A CD-AFM image of this line shows the line width is approximately 210 nm and the sidewall profile is vertical.

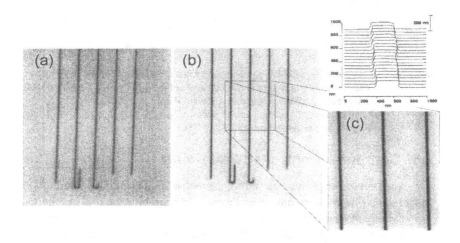

Figure 3.14:   Optical micrographs of lithographed lines before and after pattern transfer.
(a) SAL601 resist lines on polysilicon. (b) Etched polysilicon lines on oxide.
(c) High magnification of the etched polysilicon lines showing the line continuity
and uniformity, and a CD-AFM image of the etched 220-nm-wide lines showing
the vertical feature profile.

Figure 3.15 shows cross section SEM micrographs of narrow features patterned by SPL and transferred deep into the underlying silicon. A 50-nm-wide line written in SAL601 and directly etched 300 nm into the silicon is shown in Fig. 3.15(a). This line has an aspect ratio of 6:1 and a good etched profile. Fig. 3.15(b) shows a 26-nm-wide feature (measured at the top of the line) after PMMA exposure, lift-off, and a silicon etch of 260 nm. This feature has an aspect ratio of 10:1 and an excellent sidewall profile.

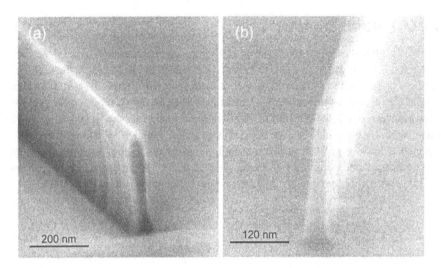

Figure 3.15: Cross-sectional SEM images of etched silicon features patterned by SPL. (a) 50-nm-wide line written in SAL601 and etched 300 nm into the silicon (6:1 aspect ratio). (b) 26-nm-wide line written in PMMA and transferred through lift-off and anisotropic etching into the silicon substrate. The etch depth is 260 nm, giving the line an aspect ratio of 10:1.

3.2.6 Computer-Controlled Patterning

We used computer software to control the path and speed of the tip during SPL. A trigger signal from the computer was used to synchronize the lithography with the tip scanning by enabling the current feedback circuit. The computer signal acted as the current setpoint (proportional to the exposure dose). We used the vector scanning strategy to write the words "MASKLESS LITHOGRAPHY" shown in Fig. 3.16. These patterns were written in PMMA positive resist and transferred through chrome evaporation, lift-off, and etching into the underlying silicon substrate. The features are approximately 36 nm wide. The vector writing strategy may be used in this way to write arbitrary patterns.

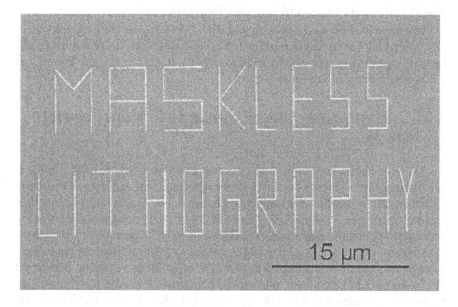

Figure 3.16: The words "MASKLESS LITHOGRAPHY" were written by SPL in PMMA positive resist using the vector scanning strategy. The resist patterns were transferred into the underlying silicon using chrome evaporation, lift-off, and etching. The etched features are 36 nm wide.

3.2.7 Summary of Contact Mode SPL

The hybrid AFM/STM system is a robust lithography tool, capable of patterning a variety of resists, substrate materials, and over topography by simply setting the desired exposure dose. The system incorporates dual independent feedback loops, one to control the tip-sample force and one to control the emission current. Constant force operation ensures that the tip follows the resist surface without exerting damaging pressure. It also minimizes the tip-to-sample spacing for improved resolution. Constant current operation delivers a fixed dose of electrons to the resist allowing reproducible and uniform lithography. The current feedback adjusts the tip-sample voltage bias to maintain a fixed emission current. This stabilizes the field-emission current and accounts for variations in resist thickness. It also facilitates line width control since the electron exposure dose (proportional to the emission current) is the critical exposure parameter. This SPL method has since been adopted by other groups because of the improved performance and reliability [33][34][35]. Simultaneous AFM and STM operation has also been used for microscopy [36]. In one example, such a system was used to characterize the structural and electrical properties of thin oxide films [37].

Tip wear is minimal in contact mode SPL. Nevertheless, a lithography technique in which the tip never touches the sample would certainly eliminate any concern regarding contamination or damage to the tip and/or sample. We present this approach in the following section.

3.3 Current-Controlled Exposures in Noncontact Mode

In the noncontact configuration, the current emitted from an AFM probe is held constant by adjusting the tip-to-sample spacing. This approach is similar to STM lithography in which the scanning tunneling microscope is operated in a field emission mode (see Section 3.1.6). The STM is generally operated under ultra high vacuum (UHV) to maintain a stable emission current [8]. Operation in UHV can be cumbersome and time-consuming. STM patterning also suffers from poor alignment capabilities since imaging may expose the resist. Here we present noncontact SPL using an AFM cantilever in air as an alternative. We found that by using an active current feedback circuit with appropriately high gain we can maintain a stable field emission of electrons from a micromachined tip in air to enable reliable lithography. The noncontact configuration should prolong the tip lifetime while retaining the high-resolution imaging capabilities of the AFM for alignment registration. It may also be extended to multiple tips operating in parallel to increase patterning throughput because of the ability to build arrays of cantilevers.

3.3.1 The Noncontact SPL System

A schematic diagram of the noncontact SPL system is shown in Fig. 3.17. A positive bias is applied to the sample to generate the field emission of electrons from the tip. The sample voltage must be large enough to ensure that the probe is out of contact with the resist surface. The vertical (Z) position of the probe is adjusted to maintain a constant current from the tip.

In our noncontact system, we mounted the AFM probe on an insulating ceramic fixture. A retaining clip on the underside of the mounting fixture made electrical contact to the probe. This contact was routed to the input of a transimpedance amplifier used to measure the emission current. The amplifier had a gain of 10^{10}, a bandwidth of approximately 2 kHz, and a current noise of about ± 2 pA. It was located close to the tip for low noise measurements. A Tektronix differential amplifier subtracted the setpoint voltage (proportional to the desired exposing current) from the measured current signal. The resulting error signal was fed into a 14 bit analog-to-digital converter (ADC) in the AFM controller. The Z

feedback was controlled by a digital signal processor (DSP) based on the error signal. Integral and proportional feedback implemented in the DSP attempted to maintain a zero error signal. A high-voltage output from the DSP was sent to the piezotube actuator to move the probe up or down.

In this mode it is necessary to use a stiff cantilever to prevent the spring-like device from snapping down to the sample as a result of the attractive electrostatic force. We used boron-doped silicon micromachined cantilevers fabricated by S. C. Minne [38] that were 73 µm long and 6 µm thick. These devices had a spring constant of approximately 340 N/m and a resonant frequency of 1.03 MHz. The integrated silicon tips had a radius of curvature of about 10 nm.

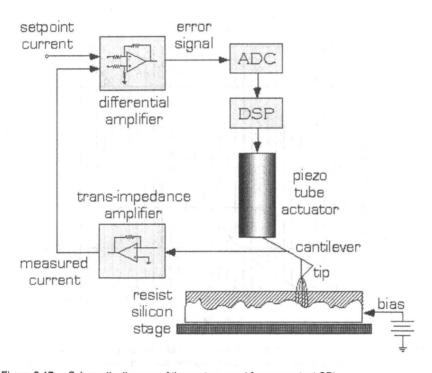

Figure 3.17:   Schematic diagram of the system used for noncontact SPL.

We spin-coated 65 nm of the negative polymer resist SAL601 on a phosphorus-doped (100) silicon substrate. The sample preparation and resist processing details are given in Section 3.1.2. The tip was initially brought into contact with the resist surface by monitoring the cantilever deflection with an optical lever sensor. Next the piezotube was fully retracted to pull the tip off the surface of the resist.

Finally, we applied a bias to the sample to extract current from the tip and enabled the current feedback. As a result, the probe moved toward the sample until the set-point current was reached.

3.3.2 Patterning in the Noncontact Mode

The relative tip height was determined from the voltage signal to the piezotube, where the scanner calibration was approximately 11 nm/V. Figure 3.18 shows the relative tip height as a function of sample bias when the emission current was maintained at 50 pA while scanning over the resist. The probe rose approximately 3.5 nm for every volt increase in the sample bias. This is consistent with the results reported by Marrian *et al.* for UHV STM resist exposure [23].

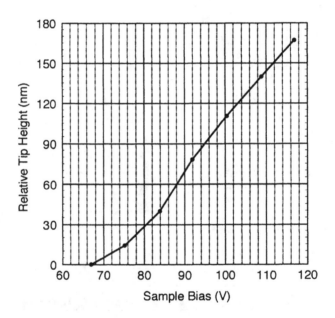

Figure 3.18: Relative tip height as a function of sample bias necessary to maintain an emission current of 50 pA while the tip was scanned over the resist surface at 10 µm/s.

The tip was scanned over the sample in a zig-zag pattern with various values of the setpoint current and sample bias. After exposure, the resist was baked, developed, and hard-baked (details are given in Section 3.1.2). The resist patterns were then imaged in the tapping AFM mode using the same cantilever that was used for lithography. Figure 3.19 shows an AFM image of patterned 65-nm-tall resist lines.

The high-resolution imaging capabilities of the AFM can also be exploited for alignment registration and profiling of latent images in the resist.

We transferred the resist features approximately 320 nm into the underlying silicon substrate using the LAM anisotropic RIE with HBr + O_2 gases (see Section 3.1.4). The Hitachi S-800 scanning electron microscope (SEM) operating at 25 keV was used to measure the width of the etched features. Uniform lines down to 28 nm in width were successfully patterned and transferred. Figures 3.20 (a) and (b) show top-down SEM micrographs of the etched lines. These lines were written with a tip scan speed of 10 μm/s and an emission current of 20 pA, giving an exposure line dose of 20 nC/cm. The sample bias was 84 V. The sample was tilted and imaged near the turn-around points. Figure 3.20(d) illustrates the high aspect ratio of the etched silicon features. The width at the corner was approximately 32 nm and the height was 320 nm.

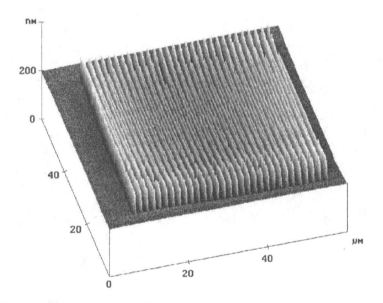

Figure 3.19: AFM image of developed SAL601 resist lines patterned by noncontact SPL. The image was taken in the tapping AFM mode with the same tip used for lithography.

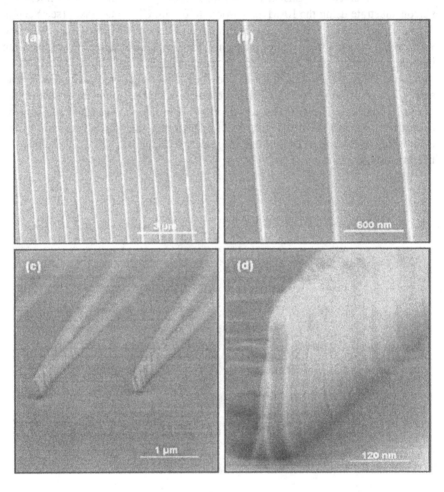

Figure 3.20: SEM micrographs of lines written using the noncontact SPL mode and etched into the underlying silicon substrate. (a) Top-down image of 28-nm-wide lines written with an exposure dose of 20 nC/cm and a sample bias of 84 V. (b) Higher magnification image showing pattern continuity and uniformity. (c) The sample was tilted to show the ends of the zig-zag patterns. (d) A higher magnification of the turn-around point. The feature is 32 nm wide at the corner and etched 320 nm deep into the silicon, yielding a 10:1 aspect ratio.

3.3.3 Line Width Control

In noncontact SPL, there are two parameters that can be independently adjusted during lithography: the emission current and the tip-sample voltage bias. The effect of the emission current on the patterned line width is shown in Fig. 3.21(a). The writing speed was 10 µm/s (therefore a current of 1 pA corresponded to an exposure line dose of 1 nC/cm). The sample bias was fixed at 84 V and the tip height was servoed to vary the emission current (or exposure dose). This mode exhibits a wide dose latitude. For example, when the dose was varied by a factor of five, the line width changed by less than a factor of two.

Figure 3.21(b) shows how the patterned line width depends on the sample voltage for noncontact SPL. As the bias was increased, the probe moved away from the sample in order to keep the emission current steady at 50 pA. Although the exposure dose was held constant, the geometrical effect of the tip moving away from the sample caused a broadening of the feature size. The line width increased from 36 nm to 79 nm when the voltage was varied from 84 V to 109 V (and from Fig. 3.18 the tip retracted about 100 nm). This demonstrates that in the noncontact mode the choice of sample bias is important for line width control. If the bias is set too high, the tip moves away from the sample and the feature dimension increases. If the bias is set too low, the tip could penetrate the resist in an attempt to reach the setpoint current. By monitoring the cantilever deflection, we can prevent this type of tip crash. There may be an optimum tip-sample bias for best resolution without touching the resist.

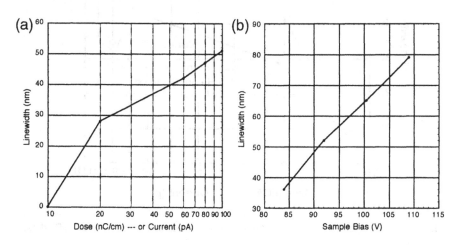

Figure 3.21: Linewidth control using the noncontact SPL mode to expose SAL601 resist.(a) Patterned line width as a function of the exposure line dose.(b) Patterned line width as a function of the sample bias.

3.3.4 Comparison of Contact and Noncontact Mode Results

In Sections 3.2 and 3.3 we showed that sub-30-nm resolution is possible using both the contact and noncontact SPL modes. In Fig. 3.22 we display the patterned line width as a function of exposure dose for both SPL modes. Recall that in the noncontact case the tip-sample bias was fixed at 84 V and the tip-to-sample spacing was servoed to vary the emission current. In the contact mode, the tip-sample spacing was minimized (10 nN contact force) and the sample bias was servoed ($\sim$ 43-46 V) to vary the emission current. The patterned lines were consistently narrower in the noncontact case than in the contact case. This seems counterintuitive, since we might expect increased beam spreading for greater tip-to-sample distances and hence degraded resolution. In Section 3.4 we explore reasons for this observation.

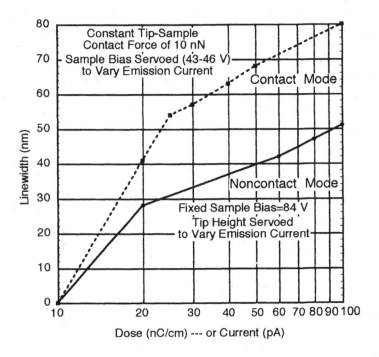

Figure 3.22: Comparison of noncontact mode (solid) and contact mode (dashed) SPL data.

3.3.5 Summary of Noncontact Mode SPL

We demonstrated that sub-30-nm features may be patterned in organic resist using noncontact mode SPL. The noncontact approach eliminates any concern over tip wear, which should improve the reliability and repeatability of SPL. Reliable noncontact SPL may facilitate the extension to parallel lithography. Simultaneous patterning with multiple tips is a direct approach to increasing the lithography throughput. A challenge for parallel lithography is to maintain independent control of the exposure by each tip. Integrated actuators used by Minne *et al.* to control the deflection of each cantilever in an array may provide a solution [39]. The integrated actuators are ZnO bimorphs located on the base of the cantilever and are used in place of the piezotube for moving the probe in the Z direction. These actuators may be used to set the exposing current from each probe by independently adjusting the tip-to-sample spacing for noncontact lithography. The sample, which would be shared by all tips, would be held at a fixed voltage. Since each tip would have independent lithography control, the system might be extended to massively parallel arrays of probes.

3.4 Simulations of Electron Field Emission and Electron Trajectories

We performed simulations of electron field emission from a biased probe tip to assist with the interpretation of the experimental results shown in Sections 3.2 and 3.3. We used the partial differential equation (PDE) tool in Matlab[a] to solve Poisson's equation and find the two-dimensional potential distribution around a biased probe tip spaced some distance from a dielectric-coated conducting plate. We chose a geometry consistent with that used in our experiments. The tip was specified to have a 10 nm radius of curvature and a 20° cone angle. We specified the region around the tip as air ($\varepsilon_r=1$) and the dielectric layer on the conducting plate as resist ($\varepsilon_r=3$[b]). We modeled both contact and noncontact mode SPL by varying the distance between the tip and resist surface for a variety of resist thicknesses.[c] The tip bias was set at 0 V and the conducting plate (analogous to the sample) was set at a positive bias. We chose a domain significantly wider than the tip size so that boundary conditions on the walls did not influence the solution around the tip.

---

a. Matlab Version 5.1 (The MathWorks, Inc., Natick, MA) run on a Sun UltraSparc.
b. Approximate permittivity of Microposit SAL601 (Shipley Corporation, Marlboro, MA).
c. Several other groups have simulated electron emission from an STM tip [44-47].

Figure 3.23 shows the potential contours in the region between the tip and sample. In Fig. 3.23(a), the tip apex is located 1 nm above a 35-nm-thick resist film; in Fig. 3.23(b), the tip apex is located 31 nm above a 35-nm-thick resist film; and in Fig. 3.23(c), the tip apex is located 1 nm above a 65-nm-thick resist film. The total tip-to-sample spacing is 66 nm in both Fig. 3.23(b) and Fig. 3.23(c). From the potential distribution, we calculated the electric field everywhere in the domain using $E = -\nabla V$. Arrows show the magnitude and direction of the electric field vector. Boundary conditions require continuity of the displacement vector across the resist-air interface. Figure 3.23 shows that the dielectric layer tries to expel the electric field, causing most of the voltage to drop between the tip and resist surface.

3.4.1   Initial Beam Size in the Contact Configuration

We extracted the electric field along the surface of the tip as a function of position and show results for the case of a tip 1 nm above a 65-nm-thick resist film [Fig. 3.24(a)]. As the voltage was increased from 20 to 50 V, the peak field (located at the tip apex) increased from 2.7 to 6.9 V/nm. The full width half maximum (FWHM) of the electric field distribution at the tip is about 14.5 nm.

Using the Fowler-Nordheim equation [Eq. (3.1)] and assuming a tip work function $\phi = 4.3$ eV for titanium, we calculated the distribution of field-emission current density from the tip. The emitted current density varies more abruptly with changes in voltage than does the electric field [Fig. 3.24(b)]. In addition, the current density distribution falls off from a maximum near the tip apex more quickly than does the field distribution [Fig. 3.24(b)]. In Fig. 3.24(c) we show a normalized plot of the current density distribution for different voltages. The initial beam diameter and beam shape do not change significantly with voltage in the "contact" (i.e., small tip-to-resist spacing) mode. In fact, the FWHM of the emitted current density distribution varies from only 2.2 to 3.4 nm over the entire range of voltages. As a result, the electron exposure dose is an appropriate parameter to describe contact mode patterning. We integrated the current density distribution to calculate the total current emitted from the tip, which ranges from about 1 pA (at 20 V) to about 6 μA (at 50 V).

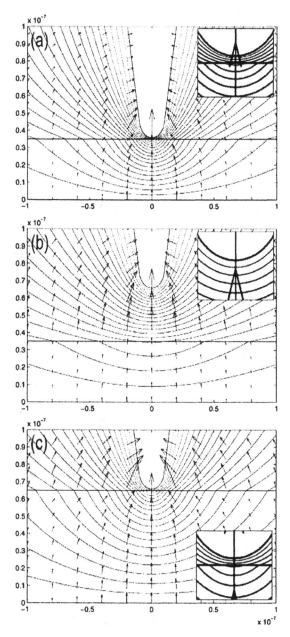

Figure 3.23:   Potential contours and electric field vectors between the tip and sample.(a) Tip 1 nm above a 35-nm-thick resist film. (b) Tip 31 nm above a 35-nm-thick resist film. (c) Tip 1 nm above a 65-nm-thick resist film.

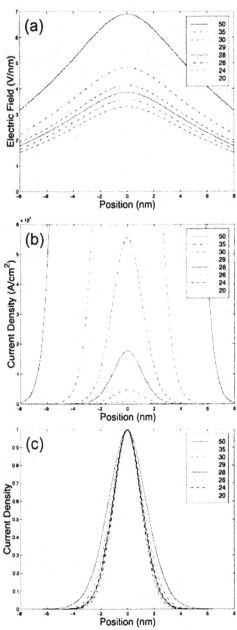

Figure 3.24: Simulation results for a tip 1 nm above the surface of a 65-nm-thick resist film. (a) Electric field distribution at the tip surface, (b) emitted current density, J, distribution, and (c) normalized J distribution for different sample biases in volts.

3.4.2 Comparison of Contact and Noncontact Configurations

As the tip is pulled away from the resist surface, there is a larger air gap (D) over which most of the voltage is dropped. This broadens the electric field distribution at the tip [Fig. 3.25] and greatly reduces the electric field strength at the tip [Fig. 3.26], even for a given tip-to-sample spacing. This can be understood by comparing Fig. 3.23(b) with Fig. 3.23(c). In the noncontact case [Fig. 3.23(b)], the potential contours are less dense between the tip and resist surface than in the contact case [Fig. 3.23(c)]. The maximum field at the tip increases linearly with sample voltage, where the slope decreases with increasing resist thickness (R) and/or D [Fig. 3.26]. The electric field within the resist is also lower in the noncontact case.

Figure 3.27 shows the normalized emitted current density distribution for several values of R and D. All curves are for a total emitted current of approximately 1 nA and exhibit a Gaussian profile. For a given D, the width and shape of the distribution do not change with R. Therefore when patterning over topography in the contact mode, where the resist thickness changes as a function of position, the initial beam shape and diameter remain constant. Yet as D increases, the distribution spreads. The current density distribution FWHM as a function of sample voltage is plotted in Fig. 3.28. The FWHM approaches 8 nm for the tip 31 nm above a 35-nm-thick resist at the highest voltage shown, where the corresponding emission current was almost 30 nA.

The total emitted current is shown as a function of voltage in Fig. 3.29. Variations in the tip-to-resist spacing cause large changes in the sample voltage required to achieve a given emission current. For instance, to achieve 1 nA emission current, a tip 1 nm above 35-nm-thick resist requires 21 V, a tip 1 nm above 65-nm-thick resist requires 27 V, and a tip 31 nm above 35-nm-thick resist requires 70 V. The electrons have significantly more energy upon entering the resist in the noncontact case than in the contact case (assuming they travel through the air gap without scattering). For a 1 nA emission current, we find from Fig. 3.23 that electrons from a tip 1 nm above 35-nm-thick resist have about 4 eV upon entering the resist, electrons from a tip 1 nm above 65-nm-thick resist also have about 4 eV, and electrons from a tip 31 nm above 35-nm-thick resist have almost 57 eV when they reach the resist.

The inset in Fig. 3.29 shows the contact mode current data on a linear scale. The shape of the curves is similar to the experimental data shown in Fig. 3.9, although the measured voltage shift from thin to thick resist was larger than that predicted by the simulations. For the simulation, we assumed an idealized spherical shape. In reality, the tip may have local asperities that can greatly change the emission properties [42]. Differences between the experimental and simulation results may be attributed to a poor estimation of the tip work function in air, of the tip shape and radius, or of the resist's dielectric constant.

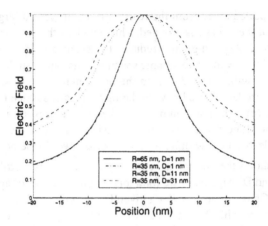

Figure 3.25: Normalized electric field distribution at the tip surface for different resist thicknesses (R) and spacings (D) between the tip apex and resist surface. All curves are for a total emitted current of approximately 1 nA.

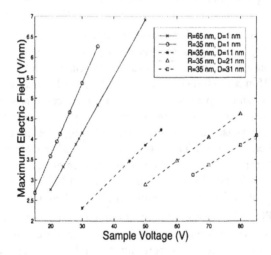

Figure 3.26: Maximum electric field at the tip apex for different resist thicknesses (R), spacings (D) between the tip apex and resist surface, and sample voltages.

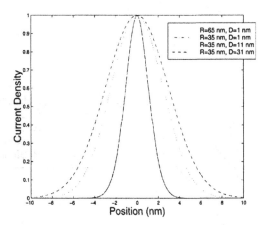

Figure 3.27: Normalized emitted current density distribution for different resist thicknesses (R) and spacings (D) between the tip apex and resist surface. All curves are for a total emitted current of approximately 1 nA.

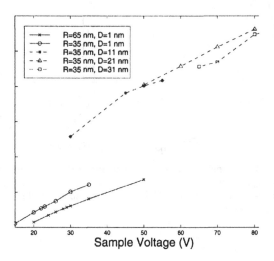

Figure 3.28: FWHM of the current density distribution for different resist thicknesses (R), spacings (D) between the tip apex and resist surface, and sample voltages.

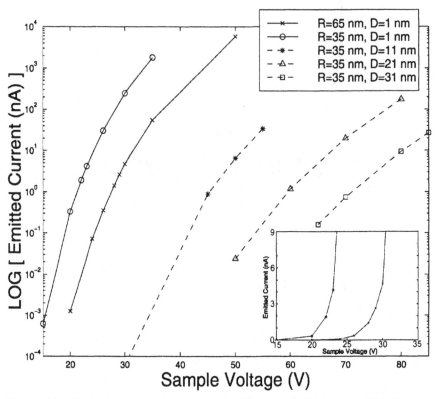

Figure 3.29: Total current emitted from the tip for different resist thicknesses (R) and spacings (D) between the tip and resist surface as a function of the sample voltage. Full graph shows data on a semilog scale. Inset shows the contact mode data on a linear scale.

3.4.3 Beam Spreading

We simulated the propagation of the field-emitted electrons from the tip through the resist to the sample, assuming they follow the electric field vectors (neglecting scattering). Figure 3.30 shows the normalized current density distribution at the tip and sample. The distribution spreads simply due to the curvature of the field lines. The amount of spreading depends on the resist thickness and the tip-resist spacing (Table 3.5), where the noncontact current distribution spreads the most from an emitted current density FWHM of 6.7 nm to a FWHM at the sample of 20.5 nm. The current density distributions at the tip and sample are approximately Gaussian in all cases. The FWHM at the sample represents the ultimate resolution of SPL for given experimental conditions in the absence of scattering. Although we expect that the low-energy electrons undergo numerous scattering events within the resist, the scattering range should be small. In Section 4.3 we show experimental data of the absorbed energy density distribution at the sample for contact mode SPL of 65-nm-thick SAL601 resist. There we measure the FWHM of the beam at the resist-sample interface to be 56 nm. It is unlikely that the difference between this value and the simulated 12.6 nm FWHM can be attributed solely to scattering in the resist. The discrepancy is likely caused by a larger initial beam diameter created, for instance, by a larger tip or different tip-resist spacing.

3.4.4 Summary of Simulation Results

Since the contact and noncontact mode cases have different beam shapes, different beam diameters, and different electron energies, the electron exposure dose may not be an appropriate parameter to use when comparing results. Therefore the data shown in Fig. 3.22 does *not* imply that noncontact SPL can achieve better resolution than contact SPL. Although we have demonstrated sub-30-nm resolution using the noncontact mode, these simulation results suggest that the contact mode is preferable to ensure high resolution, uniform patterning. The small features created by noncontact SPL may have resulted from developing conditions that generated a threshold near the narrow top of the distribution. For best resolution and reliability, the tip should be held close to the surface of a thin resist film.

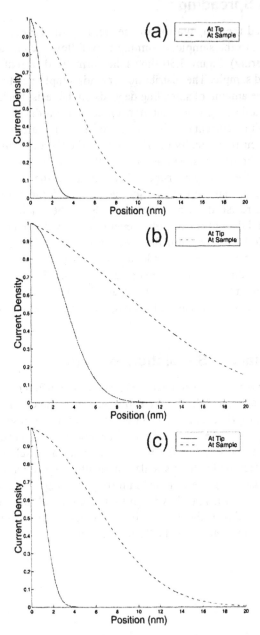

Figure 3.30: Normalized current density distribution at the tip (solid line) and sample (dashed line). (a) Tip 1 nm above a 35-nm-thick resist film. (b) Tip 31 nm above a 35-nm-thick resist film. (c) Tip 1 nm above a 65-nm-thick resist film.

R	D	Bias to emit 1 nA	Energy at Resist Surface	J FWHM at Tip	J FWHM at Sample	D FWHM
35 nm	1 nm	21 V	4 eV	2.5 nm	8.5 nm	6.0 nm
35 nm	31 nm	70 V	57 eV	6.7 nm	20.5 nm	13.8 nm
65 nm	1 nm	27 V	4 eV	2.5 nm	12.6 nm	10.1 nm

Table 3.5: Simulated electron emission FWHM and beam spreading from the tip to sample for different resist thicknesses (R) and spacings (D) between the tip and resist surface.

3.5 References

[1]   E. S. Snow and P. M. Campbell, "Fabrication of nanometer-scale side-gated silicon field effect transistors with an atomic force microscope," Appl. Phys. Lett. **66**, 1388-1390 (1995).

[2]   S. C. Minne, H. T. Soh, P. Flueckiger, and C. F. Quate, "Fabrication of 0.1 μm metal oxide semiconductor field-effect transistors with the atomic force microscope," Appl. Phys. Lett. **66**, 703-705 (1995).

[3]   K. Matsumoto, M. Ishii, K. Segawa, Y. Oka, B. J. Vartanian, and J. S. Harris, "Room temperature operation of a single electron transistor made by the scanning tunneling microscope nanoxidation process for the TiOx/Ti system," Appl. Phys. Lett. **68**, 34-36 (1996).

[4]   H. J. Kreuzer, in *Atomic and Nanometer-Scale Modification of Materials: Fundamentals and Applications,* edited by P. Avouris, NATO ASI Series, Ser. E, Appl. Sci. **239** (Boston: Kluwer Publishing, 1993), 75-86.

[5]   R. Gomer, *Field Emission and Field Ionization* (New York: American Institute of Physics, 1993).

[6]   R. H. Fowler and N. Nordheim, "Electron emission in intense electric fields," Proc. R. Soc. London Ser. A **119**, 173-181 (1928).

[7]   I. Brodie and C. A. Spindt, "Vacuum Microelectronics," Advances in Electronics and Electron Physics **83**, 1-106 (1992).

[8]   B. R. Chalamala, Y. Wei, and B. E. Gnade, "The vital vacuum," IEEE Spectrum **35**, 50 (1998).

[9]   K. A. Valiev, *The Physics of Submicron Lithography* (New York: Plenum Press, 1992).

[10]   W.-C. Wang, C.-H. Tsai, Y.-S. Fran, C.-Y. Sheu, K.-L. Tsai, and L. K. Hseu, "A high luminance FED with very low power consumption," Proc. SPIE **3421**, 69-76 (1998).

[11]   I. Y. Yang, H. Hu, L. T. Su, V. V. Wong, M. Burkhardt, E. E. Moon, J. M. Carter, D. A. Antoniadis, H. J. Smith, K. W. Rhee, and W. Chu, "High performance self-aligned sub-100 nm metal-oxide semiconductor field-effect transistors using X-Ray lithography," J. Vac. Sci. Technol. B **12**, 4051-4 (1994).

[12]   A. Claßen, S. Kuhn, J. Straka, and A. Forchel, "High voltage electron beam lithography of the resolution limits of SAL601 negative resist," Microelectron. Eng. **17**, 21-24 (1992).

[13]   F. K. Perkins, E. A. Dobisz, and C. R. K. Marrian, "Determination of acid diffusion rate in a chemically amplified resist using scanning tunnelling microscope lithography," J. Vac. Sci. Technol. B **11**, 2597-2602 (1993).

[14]   A. Broers, in *Nanostructure Physics and Fabrication,* edited by M. A. Reed and W. P. Kirk (Academic Press, San Diego, CA, 1989), p. 421.

[15]   G. H. Bernstein, D. A. Hill, and W.-P. Liu, "New high-contrast developers for poly(methyl methacrylate) resist," J. Appl. Phys. **71**, 4066-4075 (1992).

[16]   H. Liu, "Fabrication and properties of silicon nano-structures," Ph.D. Thesis, Stanford University (1995).

[17]   D. M. Tennant, L. D. Jackel, R. E. Howard, E. L. Hu, P. Grabbe, R. J. Capik, and B. S. Schneider, "Twenty-five nm features patterned with trilevel e-beam resist," J. Vac. Sci. Technol. **19**, 1304-1307 (1981).

[18]   Y. Martin and H. K. Wickramasinghe, "Method for imaging sidewalls by atomic force microscopy," Appl. Phys. Lett. **64**, 2498-2500 (1994).

[19]   K. Wilder, B. Singh, and W. H. Arnold, "Sub-0.35-μm critical dimension metrology using atomic force microscopy," Proc. SPIE **2725**, 540-554 (1996).

[20]   M. A. McCord and R. F. W. Pease, "Lift-off metallization using poly(methyl methacrylate) exposed with a scanning tunneling microscope," J. Vac. Sci. Technol. B **6**, 293-296 (1988).

References

[21]   C. R. K. Marrian, F. K. Perkins, S. L. Brandow, T. S. Koloski, E. A. Dobisz, and J. M. Calvert, "Low voltage electron beam lithography in self-assembled ultra-thin films with the scanning tunneling microscope," Appl. Phys. Lett. **64**, 390-392 (1994).

[22]   C. R. K. Marrian and E. A. Dobisz, "Scanning tunneling microscope lithography: A viable lithographic technology?" Proc. SPIE **1671**, 166-176 (1992).

[23]   C. R. K. Marrian, E. A. Dobisz, and R. J. Colton, "Lithographic studies of an e-beam resist in a vacuum scanning tunneling microscope," J. Vac. Sci. Technol. A **8**, 3563-3569 (1990).

[24]   A. Majumdar, P. I. Oden, J. P. Carrejo, L. A. Nagahara, J. J. Graham, and J. Alexander, "Nanometer-scale lithography using the atomic force microscope," Appl. Phys. Lett. **61**, 2293-2295 (1992).

[25]   M. Tortonese, R. C. Barrett, and C. F. Quate, "Atomic resolution with an atomic force microscope using piezoresistive detection," Appl. Phys. Lett. **62**, 834-836 (1993).

[26]   S. R. Manalis, S. C. Minne, A. Atalar, and C. F. Quate, "Interdigital cantilevers for atomic force microscopy," Appl. Phys. Lett. **69**, 3944-3946 (1996).

[27]   L. F. Thompson, L. E. Stillwagon, and E. M. Doerries, "Negative electron resist for direct fabrication of devices," J. Vac. Sci. Technol. **15**, 938-943 (1978).

[28]   L. E. Ocola, D. S. Fryer, G. Reynolds, A. Krasnoperova, and F. Cerrina, "Scanning force microscopy measurement of latent image topography in chemically amplified resists," Appl. Phys. Lett. **68**, 717-719 (1996).

[29]   F. K. Perkins, E. A. Dobisz, and C. R. K. Marrian, "Determination of acid diffusion rate in a chemically amplified resist with scanning tunneling microscope lithography," J. Vac. Sci. Technol. B **11**, 2597-2602 (1993).

[30]   T. Shiokawa, Y. Aoyagi, M. Shigeno, and S. Namba, "*In situ* observation and correction of resist patterns in atomic force microscope lithography," Appl. Phys. Lett. **72**, 2481-2483 (1998).

[31]   E. A. Dobisz, S. L Brandow, R. Bass, and L. M. Shirley, "Nanolithography in polymethylmethacrylate: An atomic force microscope study," J. Vac. Sci. Technol. B **16**, 3695-3700 (1998).

[32]   B. D. Terris, S. A. Rishton, H. J. Mamin, R. P. Ried, and D. Rugar, "Atomic force microscope-based data storage: Track servo and wear study," Appl. Phys. A **66**, S809-S813 (1998).

[33]   H. Sugimura and N. Nakagiri, "AFM lithography in constant current mode," Nanotechnology **8**, A15-A18 (1997).

[34]   M. Ishibashi, S. Heike, H. Kajiyama, Y. Wada, and T. Hashizume, "Characteristics of scanning-probe lithography with a current-controlled exposure system," Appl. Phys. Lett. **72**, 1581-1583 (1998).

[35]   M. Ishibashi, S. Heike, H. Kajiyama, Y. Wada, and T. Hashizume, "Characteristics of nanoscale lithography using AFM with a current-controlled exposure system," Jpn. J. Appl. Phys., Part 1, **37**, 1565-1569 (1998).

[36]   A. O. Golubok, I. D. Sapozhmikov, A. M. Solov'ev, and S. Y. Tipisev, "Scanning probe microscopy combining STM and AFM modes," Russian Microelectronics **26**, 291-296 (1997).

[37]   T. G. Ruskell, R. K. Workman, D. Chen, D. Sarid, S. Dahl, and S. Gilbert, "High resolution Fowler-Nordheim field emission maps of thin silicon oxide layers," Appl. Phys. Lett. **68**, 93-95 (1996).

[38]   S. C. Minne, "Increasing the throughput of atomic force microscopy," Ph.D. Thesis, Stanford University, 1996.

[39]   S. C. Minne, G. Yaralioglu, S. R. Manalis, J. D. Adams, J. Zesch, A. Atalar, and C. F. Quate, "Automated parallel high-speed atomic force microscopy," Appl. Phys. Lett. **72**, 2340-2342 (1998).

[40]   E. A. Dobisz, H. W. P. Koops, and F. K. Perkins, "Simulation of scanning tunneling microscope interaction with resists," Appl. Phys. Lett. **68**, 3653-3655 (1996).

[41]   E. A. Dobisz, H. W. P. Koops, F. K. Perkins, C. R. K. Marrian, and S. L. Brandow, "Three dimensional electron optical modeling of scanning tunneling microscope lithography in resists," J. Vac. Sci. Technol. B **14**, 4148-4152 (1996).

[42]   T. M. Mayer, D. P. Adams, and B. M. Marder, "Field emission characteristics of the scanning tunneling microscope for nanolithography," J. Vac. Sci. Technol. B **14**, 2438-2444 (1996).

[43]   G. Mesa, J. J. Saenz, and R. Garcia, "Current characteristics in near field emission scanning tunneling microscopes," J. Vac. Sci. Technol. B **14**, 2403-2405 (1996).

4 *SPL Linewidth Control*

The current-controlled scanning probe lithography (SPL) systems that we developed (described in Chapter 3) can reliably pattern uniform features in organic resists with dimensions below 100 nm. In this chapter, we compare electron exposures made by SPL to those made by electron beam lithography (EBL). This comparison highlights the advantages and limitations of a low-energy electron lithography technique such as SPL.

EBL is a well-established high-resolution patterning technique in which high-energy electrons are focused into a narrow beam and used to expose electron-sensitive resists. Electron beam writers are the primary tools used for patterning the masks used in photolithography. Their use in direct-write nanolithography has been limited by (1) low throughput and (2) proximity effects that hamper line width control. Low throughput results from the serial writing strategy and is a challenge for both EBL and SPL (see Chapter 8 for a discussion of parallel exposure with arrays of probes). EBL proximity effects arise from backscattered electrons whose range is considerably larger than the forward scattering range in the resist. As a result, the printed feature dimension depends on the local pattern density and size.

As integrated circuit critical dimensions (CDs) shrink, line width control becomes paramount. According to the Semiconductor Industry Association (SIA) International Technology Roadmap for Semiconductors, 100-nm line widths (projected to be in production in 2005) will require ± 7 nm CD control and 50-nm line widths (in 2011) will require ± 4 nm CD control [1]. The ability to control the narrow feature dimensions to within strict tolerances for arbitrary pattern geometries is a critical challenge for next generation lithography (NGL) tools.

SPL patterns resists with electrons field emitted from a sharp probe tip in close proximity to a sample. Low-energy SPL electrons may eliminate proximity effects and should provide other advantages over high-energy exposure. In this chapter we compare the resolution capabilities, exposure tolerances, patterning linearity, proximity effects, and exposure mechanisms of EBL and SPL.

4.1 Exposure Tools and Samples

EBL is a mature lithographic technology that was first developed in the late 1960s [2][3][4]. Figure 4.1 illustrates the components of a conventional EBL system. A high energy electron beam (typically 10-100 keV) is generated through (1) *thermionic emission,* in which a conductor is heated to give electrons sufficient energy to overcome the work function barrier; (2) *field emission,* in which a high electric field is applied to lower the potential barrier and allow electrons to tunnel out of the conductor; or (3) a combined effect known as *thermal field emission*, in which the emitter is heated and an electric field is applied [5]. The EBL system is operated under high vacuum to ensure stable emission currents. The emitted electrons pass through the electron-optical column, which contains electron lenses, beam blanking plates, a beam limiting aperture, and deflection coils. Magnetic or electrostatic deflection is used to scan the electron beam across the sample. Aberrations in the deflectors typically limit the EBL field size to a few mm^2.

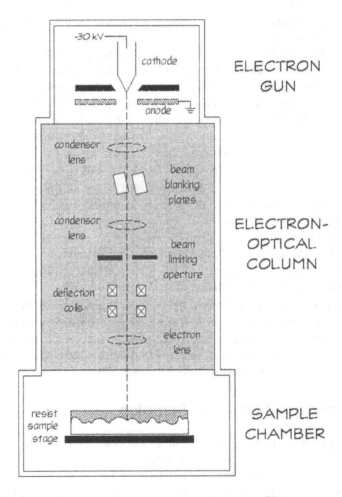

Figure 4.1: Schematic diagram of the electron beam lithography (EBL) system.

A resist-coated wafer is mounted on a stage used for sample positioning. Patterns may be written in either a *raster scanning mode* in which the beam is moved over each pixel on the sample and the beam is blanked over unwritten regions, or a *vector scanning mode* in which the beam is moved along the path to be printed. In this experiment, EBL was performed using the Hitachi HL-700F electron beam writer. It has a high brightness thermal field emission electron gun and operates at 30 keV with a 4 nA beam current.

SPL brings the electron gun into close proximity with the sample, eliminating the need for complex electron optics. A voltage bias between the tip and sample (on the order of 40-60 V for the thin resists used here) generates the field emission of electrons from the tip. We performed current-controlled SPL in the contact mode (as described in Section 3.2). The tip-sample force was held constant at 10 nN and the emission current was controlled during lithography by varying the tip-to-sample bias using custom circuitry. SPL was performed in air rather than in vacuum. A full wafer stage was used for sample positioning. The SPL field size is limited by the range of the scanner used. In our setup, the tip was mounted on a piezotube scanner with a 100 μm linear range. We used oxidation-sharpened doped silicon probe tips coated with 20 nm of evaporated titanium.

We exposed thin organic polymer resists on silicon substrates with both EBL and SPL electrons. Samples were prepared identically for EBL and SPL exposures. Starting samples were phosphorus-doped <100> 4-inch-diameter silicon wafers with 5-10 Ω-cm resistivity. Prior to resist coating, the native oxide was removed in dilute hydrofluoric acid (HF) and the wafers were singed in a convection oven at 150 °C for 30 min. Resists used for SPL must be thin to allow reasonably small tip-sample voltages (< 100 V) and to achieve high resolution. EBL also requires thin resists for ultra-high-resolution patterning, although standard EBL generally employs resists that are more than 400 nm thick. Two organic resists were used: the positive tone resist polymethylmethacrylate (PMMA) and the negative tone resist Microposit SAL601 from Shipley Company. The resists were applied using standard spin-coating techniques, which yielded pin-hole-free, uniform coatings. The sample preparation processes are described in Section 3.1.2.

Patterns consisting of lines of various widths and spacings were written over a wide range of electron exposure doses with both SPL and EBL. The developed resist patterns were transferred into the silicon substrate as described in Section 3.1.4. SAL601 resist patterns were directly etched using an anisotropic LAM reactive ion etch (RIE). PMMA resist patterns were transferred through chrome evaporation, lift-off, and RIE. The patterns were imaged using the Hitachi S-800 scanning electron microscope (SEM) operating at 25 keV. Line width measurements were made from top-down SEM micrographs of the etched silicon features taken at ≥ 200K magnification.

4.2 Sensitivity and Exposure Latitude

Figure 4.2 shows the line width versus dose data for EBL and SPL exposures of 65-nm-thick SAL601 resist. Here all features were lines written with a single pass of the electron beam or scanning probe. The exposure line dose is the incident charge per unit length (in nC/cm). The EBL emission current was held constant at ~ 4 nA while the pixel rate was varied to alter the exposure dose. For contact mode SPL, the dose was varied by adjusting either the exposing current, I, or the scan speed, s (where the ratio of I/s is the relevant line dose).

SAL601 requires a higher dose of low-energy SPL electrons than high-energy EBL electrons for exposure. In order to print a 40-nm-wide feature in SAL601, EBL requires a line dose of 0.7 nC/cm while SPL requires a dose of 20 nC/cm. The reduced sensitivity of SPL is consistent with the results of Perkins *et al.* [6]. PMMA also requires a significantly larger dose to expose with SPL than with EBL. PMMA is less sensitive than SAL601 to both 30 keV and to low-energy SPL electrons.

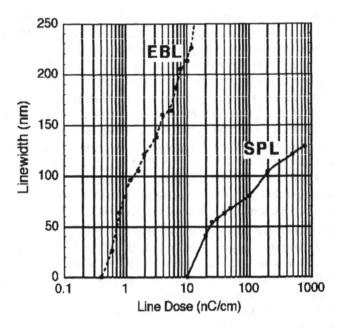

Figure 4.2: Patterned line width dependence on exposure line dose for EBL and SPL single-pass exposures of 65-nm-thick SAL601 resist on a silicon substrate.

We patterned and transferred continuous sub-50-nm features in SAL601 using both EBL and SPL. However, a more practical definition of resolution is the minimum feature size that can be achieved with an acceptable process latitude [7]. The slope of the line width versus dose curve is < 50 nm/decade for SPL and approximately 130 nm/decade for EBL. This indicates that SPL has a higher dose latitude than EBL, where dose latitude is the percent change in line width for a percent change in exposure dose. For example, when the EBL dose was increased by a factor of 10 from 0.7 to 7 nC/cm, the line width changed by a factor of 4.7 (from 40 to 187 nm). Yet when the SPL dose was increased by a factor of 10 from 20 to 200 nC/cm, the line width increased by only a factor of 2.6 (from 40 to 105 nm). The improved dose latitude means that SPL is less sensitive to dose fluctuations.

4.3 Energy Density Distribution in the Resist

We can use the line width data that we have collected to gain insight into the distribution of absorbed energy density in the resist (sometimes called the "exposure intensity distribution" [8]). The incident electrons in EBL or SPL transfer some energy to the resist that causes chemical changes in the polymer (either directly or subsequently through a post-exposure bake step). In a simple threshold approximation, we assume that there is some critical absorbed energy density in the resist above which the resist is "exposed" and below which it is "unexposed." For a negative resist such as SAL601, the unexposed regions dissolve away in the developer while the exposed areas are insoluble. At any height in the resist there is some lateral distribution of absorbed energy density resulting from the finite beam diameter and from scattering in the resist or substrate. For a given incident electron energy, the shape of this distribution will remain fixed but the magnitude will vary with incident dose. (For instance, imagine the electron beam dwelling twice as long over a given pixel. The distribution of absorbed energy would be the same, but the magnitude of the energy would be increased by a factor of two. Simulation results presented in Section 3.4 indicate this is also true of SPL even when the dose is adjusted by varying the emission current.)

We consider the absorbed energy density distribution at the resist-substrate interface resulting from a single line exposure of SAL601 resist, as depicted schematically in Fig. 4.3(a). Figure 4.3(b) illustrates the method we used to generate plots of the absorbed energy density as a function of lateral position perpendicular to the exposed line. The solid lines are representative distributions of the energy density for a line exposure at four different incident doses. The patterned line width is the width of the intersection of the distribution with the energy threshold. The threshold is such that the lowest dose shown in Fig. 4.3(b) has a maximum absorbed energy density at the threshold.

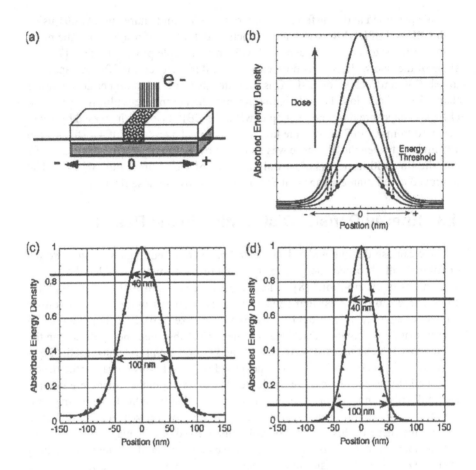

Figure 4.3: Distribution of absorbed energy density in the resist from EBL and SPL
exposures. (a) Schematic of a line exposure of resist using electrons. The energy
density distribution is determined at the resist-substrate interface perpendicular
to the exposed line. (b) Illustration of how the line width vs. dose data can be
used to generate the energy density distribution. (c), (d) Normalized absorbed
energy density as a function of lateral position from the center of the incident
beam. Data points come directly from the line width data of Fig. 4.2. (c) EBL
energy density distribution was fit with a double Gaussian. (d) SPL distribution
was fit with a single Gaussian.

A line exposed at the threshold dose would not print (line width=0). At two times this dose, the patterned line width corresponds to the width at the 50% point of the bottom energy distribution curve, as shown with the dotted lines. In general, the line width at a given dose corresponds to the width of the normalized distribution at $1/dose$. We used the line width versus dose data shown in Fig. 4.2 to plot the lateral distribution of absorbed energy density for EBL and SPL, normalizing each distribution independently.

The distribution of absorbed energy density for EBL is shown in Fig. 4.3(c). The circles correspond to data points of measured line widths from Fig. 4.2. There is an apparent tail to the distribution that we attribute to backscattered electrons. The data can be fit with the sum of two Gaussians of the form:

$$f(r) = \frac{1}{\sigma_f^2} e^{-\frac{r^2}{\sigma_f^2}} + \frac{\eta_e}{\sigma_b^2} e^{-\frac{r^2}{\sigma_b^2}} \quad , \quad (4.1)$$

where r is the distance from the center of the beam, σ_f is the half-width at the $1/e$ point for the forward scattered electrons (and in this case includes contributions from the beam diameter), and σ_b is the half-width for the backscattered electrons. η_e is the energy backscatter coefficient, representing the ratio of the total backscattered energy to the forward scattered energy deposited in the resist. For 30 keV electrons on silicon, the backscattered range is about 4 μm [9]. We fit a double Gaussian to the EBL data assuming this backscattered range. The solid line in Fig. 4.3(c) represents this normalized fit. The half-width of the narrow Gaussian is about 48 nm. This EBL data is machine specific, dependent on the electron energy and beam diameter used in the experiment. Nevertheless, the clear evidence of long-range exposure that gives rise to proximity effects is characteristic of EBL.

We performed the same procedure on the SPL line width versus dose data. The resulting energy density distribution is shown in Fig. 4.3(d). The shape of the curve can be approximated by a single Gaussian with a half-width of 33 nm.[a] There is no indication of a backscattered tail.

---

a. Li *et al.* experimentally determined the electron beam profile from an STM tip and found it was approximately Gaussian with a width comparable to that measured here [10].

4.4 Patterning Linearity Using a Pixel Writing Scheme

The lithography patterns described thus far were single-pass exposures by EBL and SPL. An alternative writing scheme uses pixels exposed at a low dose to create arbitrary-sized features. This scheme is generally used for EBL patterning today and is more attractive for patterning features of various geometries. We patterned SAL601 resist with EBL and SPL using pixels spaced by 40 nm.

Figure 4.4 shows top-down SEM images of SPL patterns written in SAL601 at a line dose of 20 nC/cm with a 40 nm pixel spacing. The line width increased linearly from 37 nm (for a one-pixel-wide line) to 239 nm (for a six-pixel-wide line). The line width as a function of incident line dose for lines of various pixel widths is plotted in Fig. 4.5. For both EBL and SPL the dose was varied by about a factor of five from the minimum dose to pattern a single-pixel-wide line. The SPL data

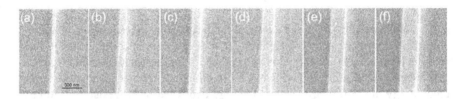

Figure 4.4: Top-down SEM images of lines patterned with SPL using the pixel writing scheme with a pixel spacing of 40 nm. The line width increases linearly from 37 nm [(a) one pixel wide] to 239 nm [(f) six pixels wide].

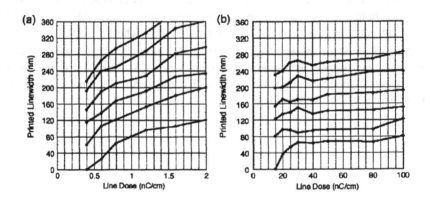

Figure 4.5: Printed line width vs. exposure dose for a pixel width of one (bottom curve) to six (top curve) for (a) EBL and (b) SPL. Pixel spacing is 40 nm.

[Fig. 4.5(b)] shows that the line width increases linearly with increases in pixel width. Also, the curves have a small slope, showing again SPL's wide exposure latitude even in the pixel writing scheme. The slope of the curves is much higher in the case of EBL [Fig. 4.5(a)]. The slope of the EBL line width versus dose curve for the six-pixel-wide feature is greater than that for the one-pixel-wide feature, showing that the linearity degrades as the dose is increased.

We display the same data in a different form in Fig. 4.6 to illustrate the issue of patterning linearity. Here the printed line width is shown versus the "target" width (or pixel width, above), where a single-pixel-wide feature has a target width of 40 nm, a two-pixel-wide feature has a target width of 80 nm, and so on. At the lowest EBL dose shown (0.4 nC/cm), the line width increased approximately linearly with pixel width, except that the single-pass line did not print [Fig. 4.6(a)]. When the dose was increased to provide the necessary exposure for the single-pass feature to print, the multiple-pass lines became considerably wider and linearity worsened. Clearly in EBL the smallest features require a higher dose to pattern at the desired line width than do the larger features. This effect, known as the "intraproximity" effect [11], is not apparent in SPL. Figure 4.6(b) shows the superior linearity achieved with SPL. For the wide range of doses shown, incremental variations in line width are possible with SPL in the pixel writing scheme.

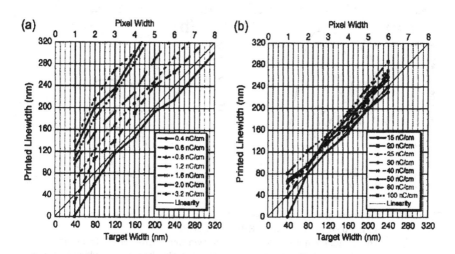

Figure 4.6: Data from Fig. 4.5 plotted to show the linearity with (a) EBL and (b) SPL. The printed line width is plotted as a function of the "target" width (or pixel width, above) for different exposure doses.

We can explain these results using our understanding of the distribution of energy density absorbed by the resist. Figure 4.7 shows the energy density absorbed as a function of lateral position for a five-pixel-wide line with 40-nm pixel spacing. The circular data points represent the line width measurements for a single pass line as plotted in Fig. 4.3. The dashed lines are the fits to the data (the double Gaussian for EBL and the single Gaussian for SPL) representing the five single line exposures separated by 40 nm. The solid line is the sum of the absorbed energy density contributed by these five pixels. The square data points represent line width measurements made on a five-pixel-wide line.

For EBL, the threshold to create a 40-nm-wide line in a single pass is about 0.85. At this threshold, a five-pixel-wide line has a width of 232 nm. It is possible to adjust the threshold or dose to create a 200-nm-wide line with five pixels, but at this threshold (or dose) no single-pass lines would have sufficient absorbed energy density to print. In the case of SPL, the energy distribution is narrower, minimizing the overlap between neighboring pixels. Figure 4.7(b) shows that at the threshold to create a 40-nm-wide line in a single pass (~0.7), the width of a five-pixel-wide line is exactly 200 nm. The confined distribution of SPL energy density allows excellent linearity and superior line width control.

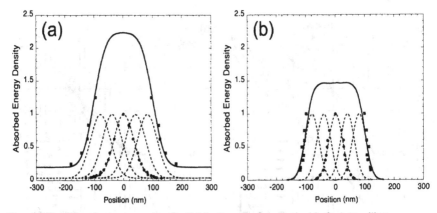

Figure 4.7: Absorbed energy density distribution of a five-pixel-wide feature with a 40-nm pixel spacing for (a) EBL and (b) SPL.
The contribution from each pixel is shown with dashed lines. The solid line represents the total energy density absorbed by the resist resulting from the five line exposures. Data points indicate line width data measured for a single-pass line and a five-pixel-wide line.

4.5 Proximity Effects

Figure 4.8 shows an SEM micrograph of a test pattern illustrating EBL proximity effects. At the top, we show lines of various pixel widths. The intraproximity effect causes the four-pixel-wide line to be more than four times the width of the one-pixel-wide line. The long-range tail of the EBL energy density distribution also contributes to "interproximity" effects, or line width variations due to the local feature density. At the bottom left of Fig. 4.8 we show five single-pass lines on a 500 nm pitch. The individual lines were resolved, but the line width was greater than the width of an isolated line. When the five lines were written on a 200 nm pitch, the lines were no longer resolved.

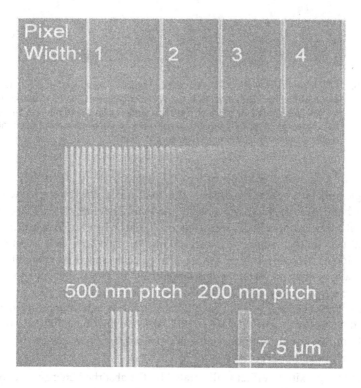

Figure 4.8:   EBL test pattern illustrating the intraproximity and interproximity effects. Proximity effects hamper EBL line width control.

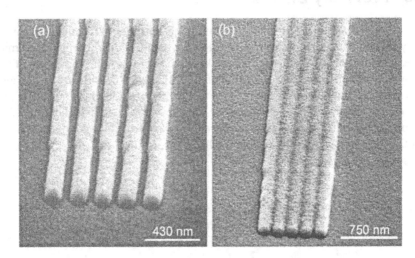

Figure 4.9: EBL interproximity effects shown for single-pass lines spaced by 200 nm. (a) Lines written with EBL at a low dose. An isolated line at this dose had a width of 64 nm. Lines on a 200-nm pitch were resolved, but the line width was 140 nm. (b) At a higher incident EBL dose, the lines were not resolved even though the isolated line width at this dose was only 120 nm.

Figure 4.9 shows an example of the EBL interproximity effect for a series of five single-pass lines spaced by 200 nm. The lines in Fig. 4.9(a) were written with a dose at which an isolated line had a width of 64 nm. The lines on the 200-nm pitch were resolved but the line width increased to 140 nm. When the dose was increased such that the isolated feature size was 120 nm, the five lines on a 200-nm pitch could no longer be resolved [Fig. 4.9(b)]. This effect becomes more significant for larger arrays of lines, as features printed relatively far away contribute to the total absorbed energy density at each line. For example, consider the 10×10 µm line gratings in the center of Fig. 4.8. The lines on a 500-nm pitch were resolved, but the line width was even greater than the width of the five lines on a 500-nm pitch. The 10×10 µm line grating with lines on a 200-nm pitch was completely washed out by long-range electron exposures.

This interproximity effect is illustrated by the absorbed energy density plots shown in Fig. 4.10. Here we show the energy density fits for single-pass lines spaced by 200 nm. The individual features are shown with dashed lines and data points show the actual line width data for the line in the center. The solid line represents the sum of the absorbed energy density, including the contributions of all features. EBL patterning of five lines spaced by 200 nm is shown in Fig. 4.10(a). Here it is clear that there is some overlap of the energy density profiles and a large tail to the distribution. For a threshold above 0.23 (corresponding to an isolated line width of < 120 nm), the lines will be resolved, while they cannot be resolved for a lower threshold (i.e., larger isolated line width). This is consistent with our experimental observations. Figure 4.10(b) shows how the absorbed energy density distribution changes when the line grating is extended 2 μm in the positive and negative directions (21 lines total). Now the individual lines can only be resolved for a threshold above 0.79, corresponding to a dose at or below that required to print an isolated line width of 48 nm. Finally, we extend the line grating 5 μm in the positive and negative directions (52 lines total). The sum of the deposited energy density [Fig. 4.10(c)] indicates that it is impossible to resolve the individual lines with a dose at which an isolated line would print.

Various proximity effect correction algorithms have been developed to achieve uniform resist exposure with EBL. For example, the exposure dose can be adjusted depending on the feature size and local feature density [12]. An alternative technique called GHOST uses a defocussed electron beam to write the inverse tone of the pattern as a second pass [13]. Others have changed the layout pattern geometry to yield exposed patterns of the desired size and shape [14]. Proximity effect correction tends to be computer-intensive and time-consuming, compounding the problems of an already-slow patterning technology. Furthermore, even complex proximity correction algorithms do not eliminate the effects completely [15]. EBL patterns shown here were exposed without proximity correction.

Proximity effect limitations are alleviated in the case of SPL. Figure 4.12 shows lines of various spacings written by SPL in SAL601 and etched into the silicon. All lines were written with a line dose of 50 nC/cm. Lines gratings are shown on a 500-nm pitch [Fig. 4.12(a)] and a 200-nm pitch [Fig. 4.12(b)]. Both gratings extend over a 10 × 10 μm area. Figure 4.12(c) shows a grid consisting of lines on a 200-nm pitch in both directions. Lines were written first in the vertical direction and next in the horizontal direction; therefore the intersections of the grid were doubly exposed. Nevertheless SPL's wide exposure latitude yields only a small change in the feature width at the corners. The printed line width was about 65 nm for all line spacings shown. We printed 200-nm-pitch grids at doses from 20 nC/cm to 80 nC/cm. All individual lines were resolved, and the line width was consistently independent of the line spacing.

There is no evidence of long-range proximity effects in SPL. We used the single Gaussian fit to the SPL absorbed energy density to show the effect of five lines on a 200-nm spacing [Fig. 4.11(a)]. The sum of the absorbed energy density (solid line) follows the individual line profiles. We extended the line grating 5 μm in both directions [52 lines total, same as the printed 10 × 10 μm grating shown in Fig. 4.12(b)] and the result is shown in Fig. 4.11(b).

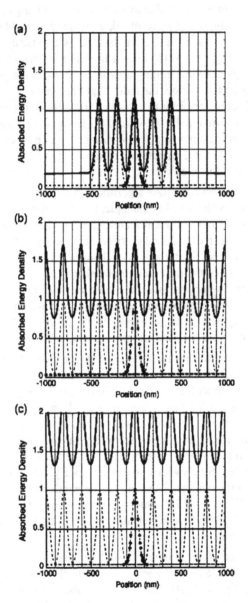

Figure 4.10: Absorbed energy density for EBL-patterned line gratings. The dashed lines represent the energy density for a single line. The solid line is the sum of the energy density contribution from all of the lines. (a) 5 lines on a 200-nm pitch. (b) 21 lines on a 200-nm pitch. (c) 52 lines on a 200-nm pitch.

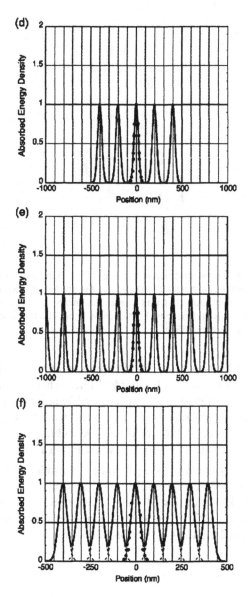

Figure 4.11:   Absorbed energy density for SPL-patterned line gratings. The dashed lines
represent the energy density for a single line. The solid line is the sum of the
energy density contribution from all of the lines. (a) 5 lines on a 200-nm pitch.
(b) 52 lines on a 200-nm pitch. (c) 9 lines on a 100-nm pitch.

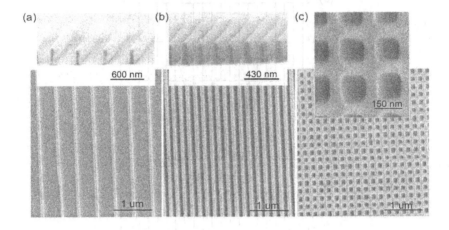

Figure 4.12:   Lines of various spacings written with SPL at a dose of 50 nC/cm in SAL601
negative resist and directly etched into the underlying silicon substrate. (a) Lines
on a 500 nm pitch. (b) Lines on a 200 nm pitch. (c) Grid with lines spaced by
200 nm in both directions. Etch depth is about 300 nm. Line width for all line
spacings is about 65 nm.

The confined SPL energy density distribution minimizes the contribution of
far-away features on the width of individual lines. There is some overlap of the
individual line profiles if the lines are spaced by 100 nm. Figure 4.11(c) shows the
absorbed energy density distribution for nine lines on a 100-nm pitch. For a threshold above 0.20 (corresponding to an isolated line width below 84 nm) it should be
possible to resolve individual lines written with SPL on a 100-nm pitch.

We have similarly found that the patterned line width is independent of feature
spacing for SPL of PMMA. Figure 4.13 shows SEM micrographs of patterns written in 50-nm-thick PMMA using SPL and transferred into the underlying silicon
using lift-off and etching. The line width is approximately 94 nm for various density patterns, from isolated lines to lines on a 200-nm pitch.

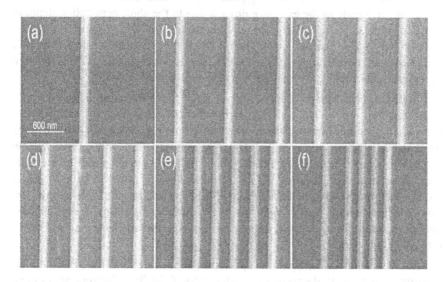

Figure 4.13: Patterns of various line spacings written with SPL in PMMA positive resist and transferred into the underlying silicon substrate using lift-off and etching. Each line was written with a single pass of the probe. SEM micrographs were taken with the sample tilted. (a) Isolated line. (b) Pitch=800 nm. (c) Pitch=650 nm. (d) Pitch=500 nm. (e) Pitch=300 nm. (f) Pitch=200 nm. The line width is 94 nm, independent of the pattern density.

4.6 Exposure Mechanisms of High- and Low-Energy Electrons

Various studies have been directed at decreasing the accelerating voltage for traditional EBL since lower energy electrons have a more confined lateral scattering range that reduces the interaction volume in the resist [16][17][18]. There are numerous challenges for creating a low-energy focused electron beam, including the beam's high sensitivity to electromagnetic fields and its low brightness [19]. Also, low-energy EBL requires thin resists because of the short penetration depth of the electrons in resist. For example, 2 keV electrons have a penetration depth of approximately 120 nm in organic resists and 1 keV electrons have a penetration depth of about 60 nm [20].

Several demonstrations of low-energy EBL on thin resists have shown that the sensitivity is improved over high-energy EBL. For example, Lee *et al.* found that the dose to expose a 100-nm feature in PMMA was a factor of 12 smaller with 2 keV electrons than with 30 keV electrons [19]. Low-energy EBL may be a more efficient process than traditional high-energy EBL since more of the low-energy electrons participate in the exposure process due to the short penetration depth in resist. In contrast, the sensitivity is *reduced* in the case of SPL, suggesting that SPL is not simply a low-energy EBL system. Lee *et al.* found that the dose tolerance increased as the accelerating voltage for EBL was reduced to 2 keV. We also found an increased dose tolerance using the very low-energy SPL electrons.

In EBL, an electron gun emits a high-energy beam of electrons a large distance away from the sample. The electron beam is then focused down to a narrow beam under high vacuum. Electrons that reach the sample may lose their energy to the resist and/or substrate. Many high-energy electrons pass through the resist without much energy loss, since the resist film is thin with respect to the electron's inelastic mean free path. (For example, 30 keV electrons travel on average > 14 µm in PMMA before they lose all of their energy [9].) Nevertheless, the high-energy electron scattering events can be modeled by a "continuous slowing down" approximation, in which it is assumed that an average electron is continuously retarded [9].

The continuous slowing down approximation is not valid for SPL since the electrons pass through the resist in a high electric field. The electrons are emitted from the tip with a very small energy and then are accelerated by the field toward the substrate. They undergo a number of inelastic scattering events in which they lose energy and then gain more from the field.

EBL exposure of SAL601 releases a photoacid in the resist. Generally no other resist chemistry takes place until the post exposure bake (PEB) initiates the cross linking reaction. In contrast, we found that significant cross linking occurs during SPL exposure. We repeated SPL single-pass exposures of SAL601 and developed without a PEB. We found no measurable changes in line width from the data shown in Fig. 4.2. Therefore we suspect either local heating (essentially an *in situ* PEB) or a fundamentally different exposure mechanism for the low-energy electrons.

It seems reasonable that some SPL electrons that scatter in the resist may not have enough energy to cause chemical changes. Multiple scattering events could heat the resist and provide the energy for the cross-linking reaction. It has been observed experimentally that at the same exposure dose, the temperature rise in the resist is higher for beams with lower accelerating voltage [21]. Nevertheless, analysis by Perkins *et al.* indicates it is unlikely that significant heating occurs in the resist during SPL exposure [6]. The insensitivity of the SPL pattern dimension to PEB time also suggests that local heating is not the cause of the cross linking during SPL exposure. Therefore, we propose that the low-energy electron exposure mechanism of SAL601 is different than that of standard EBL in that cross linking is performed directly by the incident low-energy electrons.

High-energy electrons from EBL can also directly cause a cross linking reaction and thus induce differential solubility in SAL601. The incident high-energy electrons may form a cation radical in the novolac repeat unit (oxidizing the novolac), which can lead to cross linking. At small incident doses there may not be the required threshold of cross links to change the solubility. But at high incident electron doses, significant cross linking can occur. We patterned SAL601 with 30 keV EBL and developed immediately after exposure (no PEB). We found that even at the highest incident exposure line dose used (20 nC/cm), the single pass exposures did not render the resist insoluble. But all features at least three pixels wide patterned at doses above 16 nC/cm. No patterns were evident after development (even for densely packed features) for doses below 5.6 nC/cm.

SPL electrons do not have enough energy to cause the above cationic reaction. They may, however, cause an anionic reaction by reducing the novolac (entering the lowest unoccupied molecular orbital of the novolac molecule). This can also lead to cross linking [22]. The efficiency of this exposure process may be lower than the high-energy process, which could account for the reduced sensitivity of SAL601 to the SPL electrons. It should be possible to design resists optimized for low-energy SPL exposure.

4.7 Summary

SPL exhibits a wider exposure latitude, improved linearity in the pixel writing scheme, and reduced proximity effects as compared with 30 keV EBL. However, SPL requires a higher dose of electrons for exposure than EBL. The SPL absorbed energy density in the resist can be approximated by a single Gaussian, while EBL requires a double Gaussian fit to account for the long-range tail of the distribution resulting from the contribution of backscattered electrons. SPL's confined energy density distribution allows densely packed features to be resolved. We demonstrated this by patterning 200-nm-pitch grids with SPL where all individual features were resolved. The line width of features in this grid was the same as the width of an isolated line exposed with the same dose.

4.8 References

[1]   *International Technology Roadmap for Semiconductors* (San Jose: Semiconductor Industry Association, 1997). Data also reflects 1998 update to the roadmap.

[2]   A. N. Broers, "Electron and ion probes," Proc. Symp. Electron Ion Beam Sci. Technol., 3-5 (1972).

[3]   A. N. Broers, "A new high resolution electron probe," J. Vac. Sci. Technol. **10**, 979-982 (1973).

[4]   M. Hatzakis, "Fundamental aspects of electron beam exposure of polymer resists systems," J. Electrochem. Soc. **121**, 106C (1974).

[5]   P. Rai-Choudhury, *ed., Handbook of Microlithography, Micromachining, and Microfabrication: Volume 1* (Bellingham, Wash: SPIE Optical Engineering Press, 1997).

[6]   F. K. Perkins, E. A. Dobisz, and C. R. K. Marrian, "Determination of acid diffusion rate in a chemically amplified resist with scanning tunnelling microscope lithography," J. Vac. Sci. Technol. B **11**, 2597-2602 (1993).

[7]   A. Tritchkov, R. Jonckheere, and L. Van den Hove, "Use of positive and negative chemically amplified resists in electron-beam direct-write lithography," J. Vac. Sci. Technol. B **13**, 6, 2986-2993 (1995).

[8]   K. Murata and D. F. Kyser, "Monte Carlo methods and microlithography simulation for electron and X-Ray beams," Advances in Electronics and Electron Physics 69, 175-259 (1987).

[9]   J. S. Greeneich, "Electron-beam processes," in G. R. Brewer, ed., *Electron-Beam Technology in Microelectronic Fabrication* (New York: Academic Press, 1980).

[10]   N. Li, I. Kawamoto, T. Yoshinobu, and H. Iwasaki, "Experimental measurement of the profile of the field-emitted electron beam from a scanning tunneling microscope tip," unpublished, 1999.

[11]   W. M. Moreau, *Semiconductor Lithography: Principles, Practices, and Materials* (New York: Plenum Press,1988).

[12]   T. Waas, H. Eisenmann, O Völlinger, and H. Hartmann, "Proximity correction for high CD accuracy and process tolerance," Microelectron. Eng. **27**, 179-182 (1995).

[13]   G. Owen and P. Rissman, "Proximity effect correction for electron beam lithography by equalization of background dose," J. Appl. Phys. **54**, 3573-3581 (1983).

[14]   N. Belic, H. Eisenmann, H. Hartmann, and T. Wass, "Geometrical correction of the e-beam proximity effect for raster scan systems," Proc. SPIE **3676** (1999).

[15]   M. A. McCord, "Electron beam lithography for 0.13 μm manufacturing," J. Vac. Sci. Technol. B **15**, 2125-2129 (1997).

[16]   T. J. Stark, T. M. Mayer, D. P. Griffis, and P. E. Russell, "Effects of electron energy in nanometer scale lithography," Proc. SPIE **1924**, 126-140 (1993).

References

[17] T. H. P. Chang, M. G. R. Thomson, E. Kratschner, H. S. Kim, M. L. Yu, K. Y. Lee, S. A. Rishton, B. W. Hussey, and S. Zolgharmain, "Electron beam microcolumns for nanolithography," Proc. of the Internat. Conf. on Quantum Devices and Circuits, 3-15 (1997).

[18] C. Stebler, M. Despont, U. Staufer, T. H. P. Chang, K. Y. Lee, and S. A. Rishton, "Microcolumn-based low energy e-beam writing," Microelectron. Eng. **30**, 45-48 (1996).

[19] Y.-H. Lee, R. Browning, and R. F. W. Pease, "E-beam lithography at low voltages," Proc. SPIE **1671**, 155-165 (1992).

[20] C. W. Lo, M. J. Rooks, W. K. Lo, M. Isaacson, and H. G. Craighead, "Resists and processes for 1 kV electron beam microcolumn lithography," J. Vac. Sci. Technol. B **13**, 3, 812-820 (1995).

[21] F. Murai, S. Okazaki, N. Saito, and M. Dan, "The effect of acceleration voltage on line width control with a variable-shaped electron beam system," J. Vac. Sci. Technol. B **5**, 105-109 (1987).

[22] D. Tully, A. Trimble, and J. M. J. Fréchet, Department of Chemistry, University of California, Berkeley, California, private communication.

5   Critical Dimension Patterning Using SPL

Transistor gate patterning is the primary application of a high-resolution lithographic system in the semiconductor industry. The gate itself is typically a long, narrow line of polysilicon whose width (known as the transistor gate "length") determines the device switching speed. The uniformity of the gate is critical for device electrical performance and yield. Gate patterning is performed after significant device processing. Therefore the feature must be accurately aligned to the previously patterned regions. It must also be written over the sample topography created by the prior fabrication steps.

In this chapter we demonstrate critical dimension patterning using scanning probe lithography (SPL). We used a "mix and match" lithography approach to create p-channel metal-oxide-semiconductor field-effect transistors (pMOSFETs). Only the gate was patterned using SPL; the remaining five lithography levels were patterned using photolithography. At the time of this research (1996), the gate length in production MOSFETs was about 250 nm. We targeted the pMOSFET physical gate length at 100 nm. Several other researchers had investigated 100 nm pMOSFETs and found improved device performance over longer-channel devices [1-5].

5.1   100 nm pMOSFET Device Fabrication

The main process steps for the pMOSFET fabrication are illustrated in Fig. 5.1. Starting samples were n-type <100> 4-inch-diameter silicon wafers with 5-10 Ω-cm resistivity. We performed semirecessed local oxidation of silicon (LOCOS) to electrically isolate neighboring devices [6]. During LOCOS we formed a thick field oxide (thickness $\approx$ 450 nm), which created a step of approximately 200 nm between the field and active areas. We implanted arsenic ions (energy=100 keV, dose=10^{13} cm^{-2}) through a thin oxide to dope the channel. The implant oxide was then stripped and a 57 Å gate oxide was thermally grown. Next 100 nm of polysilicon was deposited. We implanted the polysilicon with BF$_2$ ions (energy=10 keV, dose=10^{15} cm^{-2}) and performed a 10 s rapid thermal anneal (RTA) to electrically activate the dopants since the polysilicon must be conductive for SPL. A 50-nm-thick film of low temperature oxide (LTO) was deposited on top of the gate stack. The LTO was patterned by photolithography and etched in dilute HF to form the gate contact pads. Next we patterned the gate using SPL (see Section 5.2).

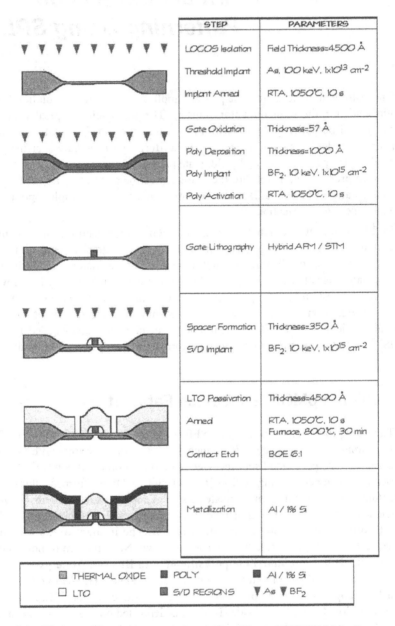

STEP	PARAMETERS
LOCOS Isolation	Field Thickness=4500 Å
Threshold Implant	As, 100 keV, 1x10^{13} cm^{-2}
Implant Anneal	RTA, 1050°C, 10 s
Gate Oxidation	Thickness=57 Å
Poly Deposition	Thickness=1000 Å
Poly Implant	BF$_2$, 10 keV, 1x10^{15} cm^{-2}
Poly Activation	RTA, 1050°C, 10 s
Gate Lithography	Hybrid AFM / STM
Spacer Formation	Thickness=350 Å
S/D Implant	BF$_2$, 10 keV, 1x10^{15} cm^{-2}
LTO Passivation	Thickness=4500 Å
Anneal	RTA, 1050°C, 10 s Furnace, 800°C, 30 min
Contact Etch	BOE 6:1
Metallization	Al / 1% Si

THERMAL OXIDE	POLY	Al / 1% Si
LTO	S/D REGIONS	▼ As ▼ BF$_2$

Figure 5.1: Diagram of the main process steps in the 100 nm pMOSFET fabrication.

After gate patterning and polysilicon etching, a 35 nm oxide spacer was formed by LTO deposition followed by anisotropic reactive ion etching (RIE) [7]. Shallow source/drain junctions were created by low-energy BF_2 ion implantation (energy=10 keV, dose=10^{15} cm^{-2}). We passivated the entire structure with LTO and then performed an RTA to activate the dopants. A furnace anneal was performed before contact formation and metallization. Finally, the wafers were annealed in a forming gas at 400 °C.

5.2   Gate Level Lithography Using SPL

The transistor structure after polysilicon deposition and gate pad formation is shown schematically in Fig. 5.2(a). We spun SAL601 negative-tone resist diluted in Microposit Thinner Type A[a] on top of this structure [Fig. 5.2(b)]. On a flat sample, this solution yielded a nominal resist thickness of about 65 nm, but on these wafers the resist thickness varied significantly as a function of position due to the sample topography. Figure 5.3(a) shows the polysilicon and resist topography as profiled by the AFM. The resist, which was spun as a liquid, tried to planarize the surface but did a poor job because of the large sample steps. The resulting resist thickness variations are shown in Fig. 5.3(b).

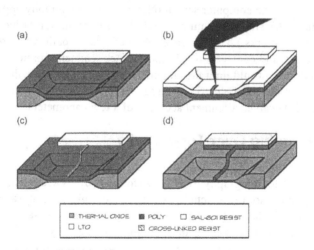

Figure 5.2:   Schematic diagrams of the 100 nm pMOSFET gate patterning by SPL.
(a) LTO gate pad was photolithographically defined on polysilicon; (b) SAL601was spun on the sample and SPL was performed; (c) Exposed gate remained after development; (d) Polysilicon was etched by RIE and the resist was stripped.

a. SAL601:Thinner A ratio 1:3 spun at 7 krpm for 30 s.

A titanium-coated silicon tip was used as the electron emitter. Gate lithography was performed in air using current-controlled SPL in the contact mode (also called hybrid AFM/STM lithography and described in Section 3.2). The exposure dose was controlled using a real-time feedback system that adjusts the tip-sample voltage to maintain a constant exposing current. The force between the tip and the resist surface was independently controlled during exposure and maintained at about 10 nN. The tip was moved over the sample topography from the field region, into the active area, back onto the field region, and onto the gate pad [Fig. 5.2(b)].

5.2.1 Overlay Registration

For a "mix and match" lithography approach to be successful, there must be an accurate method for aligning new features to patterns previously written on the sample by another lithography technology. It is expected that 100-nm technology will require an overlay registration tolerance of 35 nm [8]. To achieve accurate alignment using SPL, we exploited the high-resolution imaging capabilities of the AFM.

Prior to gate patterning, we imaged the transistor structure with the AFM to precisely determine the location for the gate. This image was imported into custom lithography software that controls the path and speed of the tip. The location of the gate was drawn on the computer screen relative to the topography apparent in the imported image. (A similar technique could be used to align relative to pre-patterned alignment marks.) The tip traced the designated path with preset speed, exposing current, and contact force. In most cases, scanner nonlinearities such as hysteresis and creep would limit the alignment accuracy. However, the closed loop feedback used in our piezoelectric tube scanner (ScanMaster™ from Park Scientific Instruments) enabled alignment accuracy of a few nanometers.

5.2.2 Patterning Over Topography

LOCOS isolation creates a topographic step between the field and active area over which the gate lithography must be performed [6]. LOCOS was used extensively in semiconductor manufacturing until very recently when it was replaced by a shallow trench isolation (STI) scheme. STI yields smaller sample steps than LOCOS, yet there remains some gate level sample topography.

Patterning over topography is a challenge for most forms of lithography. Resist spun on a wafer with topography tends to planarize, resulting in resist thickness variations across the sample. In photolithography, this causes line width variations across the wafer due to the scattering of light off the topography and to the interference effect [9]. The resist thickness may in fact exceed the system's depth of focus in certain regions, particularly for high-resolution lithography. It is also difficult to achieve uniform patterns over topography using electron beam lithography (EBL).

This is a result of electrons backscattering from the substrate and changing the feature profile near a topographic step [10][11].

For different reasons, it is impractical to pattern a wafer with topography with either the STM or the AFM. In STM lithography the problem is usually one of beam spreading causing line width variations between regions. A more aggressive attempt to limit beam spreading can cause the tip to penetrate the resist in the thicker regions. With the AFM, the chosen voltage bias will likely pattern only one region (perhaps the thinner resist), but not expose another region at all.

The hybrid AFM/STM lithography system was designed to circumvent these problems. This system delivers a constant dose of electrons to the resist while maintaining a minimum tip-sample spacing (see Section 3.2). This allows us to pattern continuous and uniform features over topography. Here the SAL601 resist thickness changed abruptly from less than 30 nm to more than 90 nm over the distance of only a micron at the field-to-active transition [Fig. 5.3(b)]. Using the hybrid AFM/STM system we patterned continuous lines over this topography in a single pass.[a] Figure 5.3(c) shows the current and voltage data taken while writing over the topography. The applied bias ranged from below 22 V to more than 81 V in order to keep the current constant at 0.05 nA.

After exposure, the wafer was given the post exposure bake (PEB) on a hot plate (1 min, 115 °C) and developed in MF-322 for 10 min, leaving the exposed resist line over topography [Fig. 5.2(c)]. Next the polysilicon was anisotropically etched in an SF_6 + Freon115 plasma. During the etch, the patterned SAL601 resist masked the gate and the photolithographically-defined LTO pad masked the gate contact pad [Fig. 5.2(d)]. Finally, the remaining SAL601 resist was stripped leaving a continuous polysilicon line over the oxide topography and connected to the gate pad. Figure 5.4 shows a 100-nm-wide etched polysilicon feature across the transition from the field region to the active area. The image, taken by the critical dimension AFM (CD-AFM), shows the line is uniform over the step.

---

a. The speed at which the current can be held constant during patterning over topography is limited by the tip-sample capacitance. Speed constraints are discussed in Section 6.1.

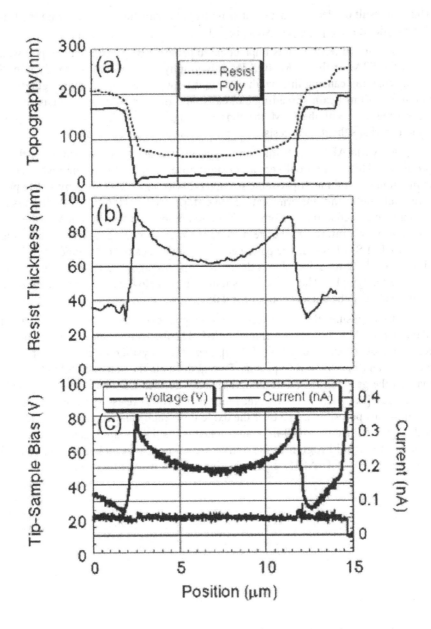

Figure 5.3: (a) Polysilicon and resist topography created by LOCOS isolation.
(b) Resist thickness variation as a function of position. (c) Current and
voltage during current-controlled lithography over the topography.

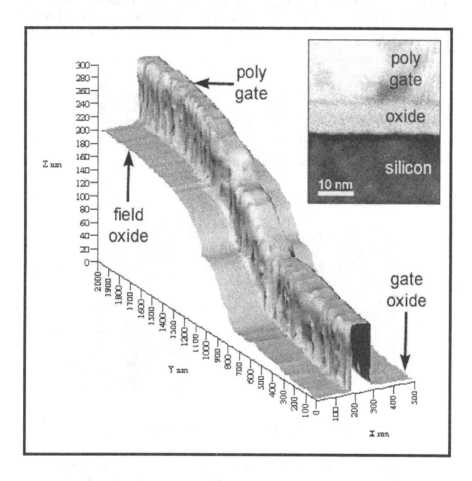

Figure 5.4: CD-AFM image of an etched 100-nm polysilicon gate over 180 nm of topography.The topography was created by LOCOS isolation. The line was patterned by SPL of SAL601 organic resist at a scan speed of 1 μm/s and with an exposure line dose of 500 nC/cm. *Inset:* TEM micrograph of the gate stack. The gate oxide in the final devices was 57 Å thick.

5.3 PMOSFET Device Characteristics

We fabricated more than 50 pMOSFETs using SPL for gate level patterning. We varied the SPL exposure dose to achieve physical gate lengths from 61 nm to 170 nm. Figure 5.5 shows an optical micrograph of a completed device. The gate, patterned by SPL, can be seen crossing the active area between the source and drain contacts. There are four electrical leads per device: source, drain, gate, and substrate.

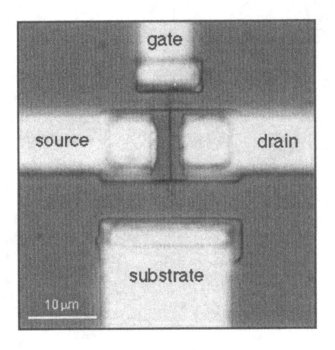

Figure 5.5:   Optical micrograph of a completed 100 nm pMOSFET.

We tested the devices at an electrical probe station using a HP4155A parameter analyzer. The electrical characteristics of a pMOSFET with a physical gate length of 130 nm, an effective channel length (L_{eff}) of 100 nm, and a gate width of 10 µm are shown in Fig. 5.6. A saturation current drive of 0.244 mA/µm and saturated transconductance (g_m) of 154 mS/mm were achieved at a –2 V power supply. The threshold voltage (V_t) was –0.41 V. Devices with smaller gate lengths showed excessive leakage, which we attribute to drain-induced barrier lowering (DIBL). DIBL is a short-channel effect in which the drain voltage affects the potential distribution under the gate, resulting in increased subthreshold conduction [12][13].

The threshold voltage rolloff, maximum saturated current (I_{dmax}), and g_m as a function of effective channel length are plotted in Fig. 5.7. Transistors with gate lengths larger than 170 nm were patterned with photolithography to show the trend as a function of gate length. The sharp rise in current drive and transconductance with decreased channel length is the impetus for MOSFET scaling to still smaller dimensions.

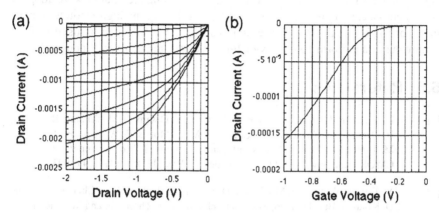

Figure 5.6: Current-voltage characteristics of a pMOSFET with L_{eff}=100 nm.
(a) I_d versus V_d where V_g was varied from 0 V to –2 V in steps of –0.25 V.
(b) I_d versus V_g at V_d=–0.1 V showing that V_t=–0.41 V.

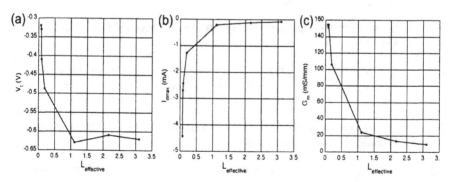

Figure 5.7: Electrical characteristics as a function of L_{eff} for 10-μm-wide pMOSFETs.
(a) Threshold voltage (V_t) roll-off. (b) Maximum saturated current, I_{dsat}.
(c) Maximum saturated transconductance, g_m.

5.4 Summary of "Mix and Match" Lithography

We demonstrated the compatibility of SPL with standard semiconductor processing by fabricating pMOSFETs with 100 nm channel lengths. We used a "mix and match" lithography approach in which SPL was used only for critical dimension gate patterning while conventional photolithography was used for all other layers. Standard resist coating, developing, and etching techniques were adopted. Nanometer-scale alignment was achieved by exploiting the high-resolution imaging capabilities of the AFM. We demonstrated that SPL can be used to form continuous and uniform 100-nm-wide lines over 200 nm of sample topography.

5.5 References

[1]   K. F. Lee, R. H. Yan, D. Y. Jeon, G. M. Chin, Y. O Kim, D. M. Tennant, B. Razavi, H. D. Lin, Y. G. Wey, E. H. Westerwick, M. D. Morris, R. W. Johnson, T. M. Liu, M. Tarsia, M. Cerullo, R. G. Swartz, and A. Ourmazd, "Room temperature 0.1 μm CMOS technology with 11.8 ps gate delay," Proc. IEDM, 131-134 (1993).

[2]   Y. Taur, S. Wind, Y. J. Mii, Y. Lii, D. Moy, K. A. Jenkins, C. L. Chen, P. J. Coane, D. Klaus, J. Bucchignano, M. Rosenfield, M. G. R. Thomson, and M. Polcari, "High performance 0.1 μm CMOS devices with 1.5 V power supply," Proc. IEDM, 127-130 (1993).

[3]   T. Hori, "A 0.1 μm CMOS technology with tilt-implanted punchthrough stopper," Proc. IEDM, 75-78 (1994).

[4]   B. Davari, "CMOS technology scaling, 0.1 μm and beyond," Proc. IEDM, 555-558 (1996).

[5]   S. C. Minne, H. T. Soh, P. Flueckiger, and C. F. Quate, "Fabrication of 0.1 μm metal oxide semiconductor field-effect transistors with the atomic force microscope," Appl. Phys. Lett. **66**, 703-705 (1995).

[6]   E. Kooi, *The Invention of LOCOS* (New York: The Institute of Electrical and Electronics Engineers, Inc., 1991).

[7]   S. Wolf, *Silicon Processing for the VLSI Era* (Sunset Beach, California: Lattice Press, 1990).

[8]   *International Technology Roadmap for Semiconductors* (San Jose: Semiconductor Industry Association, 1997). Data also reflects 1998 update to the roadmap.

[9]   W. H. Arnold, B. Singh, and K. Phan, "Line width metrology requirement for submicron lithography," Solid State Technol. **32**, 139-145 (1989).

[10]   L. Bauch, U. Jagdhold, and M. Bottcher, "Electron beam lithography over topography," Microelectron. Eng. **30**, 53-56 (1996).

[11]   T. Waas, E. Eisenmann, O. Vollinger, and H. Hartmann, "Proximity correction for high CD accuracy and process tolerance," Microelectron. Eng. **27**, 179-182 (1995).

[12]   J. M. Pimbley and J. D. Meindl, "MOSFET scaling limits determined by subthreshold conduction," Trans. Elec. Dev. **36**, 1711-1721 (1989).

[13]   R. R. Troutman, "VLSI limitations from drain-induced barrier lowering," IEEE J. of Solid-State Circuits **SC-14**, 383-391 (1979).

6   *High Speed Resist Exposure With a Single Tip*

We have shown that current-controlled scanning probe lithography (SPL) can reliably pattern nanometer-scale features in resist. However, the serial nature of SPL makes it much slower than mask-based techniques such as photolithography, x-ray lithography, or extreme ultraviolet lithography. An advantage of a direct write approach is that it does not require expensive and time-consuming mask fabrication. SPL may also have superior alignment capabilities. Nevertheless, in order for SPL to become a viable technology for high-resolution semiconductor lithography, the throughput must be dramatically increased.

6.1 High Speed Patterning of Siloxane SOG

To date, the writing speed of SPL has been limited to about 1-10 μm/s for both electric-field-enhanced oxidation and the exposure of various organic, inorganic, and monolayer resists. This makes patterning of large areas with reasonable throughput impractical. The speed limitations of exposing electron sensitive films had not previously been thoroughly investigated. Since the electron dose can be made almost arbitrarily high using a scanning probe (electron dose depends strongly on the bias voltage), we considered it likely that electron-sensitive resists could be patterned at speeds faster than previously demonstrated. In this section we report on high speed patterning of siloxane spin-on-glass (SOG).

6.1.1 Mechanism of Exposure

A schematic of the siloxane SOG molecular structure of is shown in Fig. 6.1. The siloxane type of SOG consists of a methyl group (CH_3) and silanol groups (Si-OH) bonded to the Si atoms in the Si-O backbone. The organic content refers to the number of methyl groups, which is proportional to the number of Si-C bonds. SOG with high organic content has been used as a planarization layer in the IC industry due to its attractive mechanical properties such as low deposition stress, high resistance to cracking, and low film shrinkage [1]. The silanol group with its high polarization (Si-O-H+) is a site for absorption of water. Thus SOG readily absorbs water molecules that must be removed during the processing.

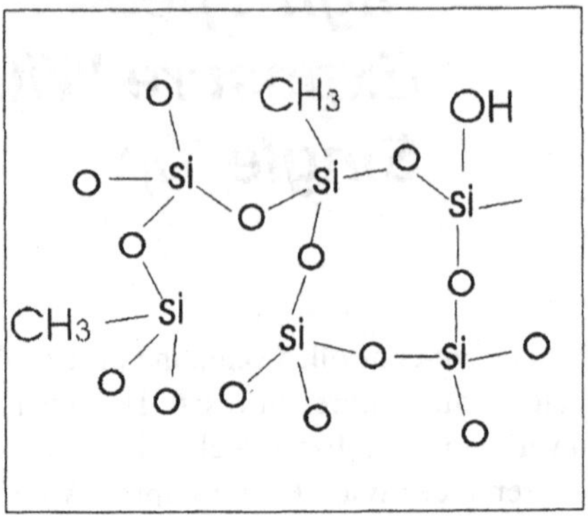

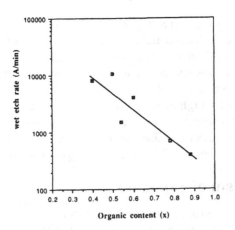

Figure 6.1: Schematic of the siloxane SOG chemical structure. Si-CH3 (siloxane) and Si-OH (silanol) are present. The highly polarized (Si-O-H+) bonds will absorb water molecules which are not shown in the schematic.

Figure 6.2: Wet etch rate as a function of organic content. As the organic content decreases, the wet etch rate increases exponentially.

It is believed when the SOG film is exposed by electrons from the tip, the methyl group is decomposed, thereby decreasing the organic content. The resulting dangling silicon bonds are filled with the silanol group. This is consistent with the results of Nakano *et al.* [2]. As the silanol group occupies the dangling silicon bonds, the etch rate of SOG in buffered oxide etch (BOE) decreases exponentially with increasing organic content as shown in Fig. 6.2. The large differential etch rate between the exposed and unexposed regions makes SOG an effective positive tone resist material for SPL.

6.1.2 Experimental Procedure

We used methylsiloxane SOG (ACCUGLASS-111, Allied Signal, Inc.) with 11% organic content. The sample substrate was a 2.5 Ω-cm, n-type Si <100> wafer. The SOG was spun onto the wafer at 5,000 rpm for 20 s which yielded a thickness of 100 nm. The sample was heated to 280 °C for 5 min, and then mounted in the AFM (Autoprobe CP, Park Scientific Instruments) operating in air in the contact mode. A positive voltage was applied to the sample to induce field emission from the tip. The emitted current was measured with a I-V converting amplifier. A d.c. voltage of 70-100 V was applied to the substrate while monitoring the current. The measured current was in the range of 0.5-3.0 nA. The applied voltage and measured current roughly followed the Fowler-Nordheim behavior (Fig. 6.3) whose characteristics are observed by measuring a straight line when $log\ (I/V^2)$ is plotted as a function of I/V. A more detailed analysis of this behavior is discussed in Section 3.2.2.

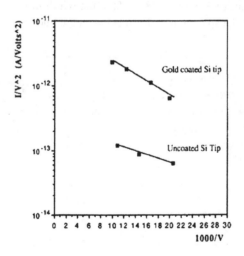

Figure 6.3:   Measured emission current vs. applied voltage. The emission characteristics show Fowler-Nordheim behavior

After the current settled to the desired value, the tip was scanned over the sample across a 90 x 90 μm area at a predetermined scan speed. The scan speeds were varied from 180-3024 μm/s. Larger scan areas and higher scan speeds were limited by the response of the piezotube scanner. After the exposure, the sample was etched in 100:1 BOE for 15 s, rinsed in de-ionized water, and blown dry with nitrogen.

We used both doped silicon tips and silicon tips coated with 30 nm of gold for patterning. The gold coated tips showed excessive wear. The uncoated silicon tips showed less wear and were used primarily in the experiments. The results shown here are from uncoated doped silicon tips.

6.1.3 Results of SOG Patterning

No latent image appears in the SOG surface if the film is imaged with the AFM immediately after exposure and before development. The exposed SOG was developed in 100:1 BOE. In this etch, the unexposed region had an etch rate of 3.6 Å/s. The etch selectivity was 20:1 between the exposed and unexposed regions. AFM micrographs of developed patterns are shown in Fig. 6.4.

The depressions (darker contrast) correspond to the regions of the SOG scanned by the tip. The SOG in these regions was etched away during the development step. In Fig. 6.4(a) we show an AFM micrograph of fine lines on a 200 nm spacing patterned with a current of 0.8 nA and a scan speed of 180 μm/s. The line width is 40 nm. In Fig. 6.4(b), we show the structure written with a scan speed of 1 mm/sec and a current of 2.0 nA at 78 V. The line width in the image is approximately 100 nm with the spacing of 1.4 μm between the lines.

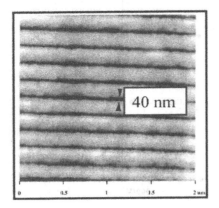 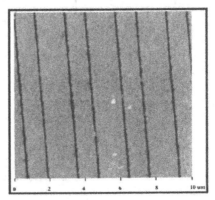

Figure 6.4: Results of SPL patterning of SOG. (a) SOG pattern of a fine line at exposed at 0.8 nA and 180 μm/s. (b) High speed patterning at 2.0 nA and 1 mm/sec.

6.1.4 Discussion

Using field emitted electrons we were able to achieve about a three order of magnitude improvement in the patterning scan speeds compared to the method of electric-field-enhanced oxidation discussed in Chapter 2. This is not surprising because the mechanism of patterning is not reaction limited at the surface, but is instead limited by the amount of current that can be emitted by the tip. The field emitted current has a large range that can be controlled by a small change in the voltage bias. We exposed siloxane SOG at scan speeds in excess of 3 mm/s, and the limitation came from the frequency response of the piezotube scanner.

However, we found that it is generally difficult to get reliable and repeatable control of the line widths in the constant voltage mode with the AFM. The difficulty arises from the fact that the field emission current is an extremely sensitive function of the film thickness, the bias voltage, and the tip shape. A few nm deviation in film thickness from sample to sample, slight deviation in the bias voltage setting, or use of a different tip produces variations in line width.

6.2 Current-Controlled SPL at High Speeds

In order to maintain the reliability of the SPL resist exposure systems described in Sections 3.2 and 3.3, we need to provide real-time control of the exposing current during lithography. In Chapter 4 we showed that the electron exposure dose—the ratio of exposing current to writing speed—is the critical parameter for patterning organic resists. The pattern dimension can be set with the appropriate choice of the exposure dose. To pattern at high speeds, we must increase the exposing current proportional to the speed increase. For example, to achieve a 100 nC/cm exposure dose (typical for SAL601 resist) at a 10 μm/s scan speed, we used an exposing current of 100 pA. To maintain the same dose while writing at 1 mm/s, we require 10 nA current emission from the tip. At a 10 cm/s scan speed, we would need 1 μA. Since the current depends strongly on the bias voltage (see Section 3.2.2), we can increase the emission current dramatically with only a few volts change in the tip-sample bias. Field-emission currents in excess of 500 μA have been measured from sharp tips in vacuum [3]. Current control ensures good uniformity and repeatability of the patterned features. The lithographic speed is limited primarily by the bandwidth of the feedback loops used to control the tip-sample force, spacing, and current during patterning.

6.2.1 Control of the Tip-Sample Force or Spacing at High Speeds

The mechanical response of the actuator that moves the probe up and down limits the scan speed with constant force maintained. Generally, the piezotube scanner is used as the actuator. This large device typically has a resonance below 1 kHz, limiting scan speeds to below 200 μm/s. Manalis *et al.* demonstrated that the tip velocity can be increased by at least an order of magnitude by using a zinc oxide (ZnO) piezoelectric actuator integrated onto the cantilever [4]. Minne *et al.* used such cantilevers for high speed imaging with multiple tips where the tip-sample force was maintained simultaneously and independently by each cantilever [5][6].

Integrated actuators may provide a solution for high speed SPL in the noncontact mode. Noncontact current-controlled SPL was described in Section 3.3. This system adjusts the tip-to-sample spacing to maintain a constant emission current during exposure. The speed of noncontact SPL is limited by the response of the actuator used to move the probe up or down. By replacing the piezotube actuator with an integrated ZnO actuator, we expect to achieve real-time control of the exposure dose at high speeds.

The ZnO actuator could also be used to maintain a constant tip-sample force during high-speed contact mode SPL (also known as hybrid AFM/STM SPL, see Section 3.2). As a simpler alternative, we investigated the feasibility of performing exposure lithography in the constant-height AFM mode, where the tip is scanned in contact with the sample without controlling the tip-sample force. Minne *et al.* found that the quality of oxidation lithography was superior when constant force was maintained. The higher forces between the tip and hard silicon sample apparently damaged the tip and degraded patterning fidelity [7]. Because the surface of an organic resist is soft and pliable, we do not expect small variations in the applied force to damage the tip. However, excessive force between the tip and sample could cause the tip to penetrate (or scratch) the resist.

We tested exposure lithography with and without real-time force feedback. For constant-height scanning, we lowered the tip toward the resist-coated sample until the cantilever was deflected slightly (~10 nN force between the tip and resist). The tip was then moved in the *x-y* plane of the sample and the current feedback was enabled. Figure 6.5 shows that lines written with and without force feedback appear equally good. In fact, in some instances patterns written in the constant-height mode had superior uniformity. The tip-sample bias used to generate the electron beam contributes an electrostatic force that has an adverse effect on the force feedback. We observed this effect as a variation in voltage during lithography on flat samples; the voltage is generally more steady in the constant-height mode. If the cantilevers were sufficiently compliant, this constant-height scanning scheme should also work for patterning over topography. Therefore either by incorporating integrated actuators or by operating in the constant height AFM mode, the lithography speed is not limited by the response of the force feedback. Maintaining the emission current (or exposure dose) at a fixed level during lithography at high scan speeds is the remaining challenge for contact mode SPL.

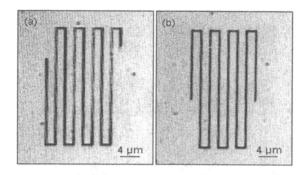

Figure 6.5:   Optical micrographs of developed SAL601 resist lines patterned with SPL at a speed of 10 μm/s and with a constant exposing current of 0.5 nA. (a) Lines written with a constant tip-to-sample force of 10 nN maintained during lithography. (b) Lines written in the constant-height AFM mode.

6.2.2 Control of the Emission Current at High Speeds

The hybrid AFM/STM lithography system varies the tip-sample voltage bias (HV) to maintain a fixed emission current. The effective bandwidth of the current feedback is limited by the presence of a finite tip-sample capacitance, C_{t-s}. A change in voltage generates a displacement current proportional to C_{t-s}. Therefore the total measured current (I_{meas})—which the current feedback tries to keep constant—is a sum of the exposing current through the resist (I_{res}) and this capacitive current (I_{cap}):

$$I_{meas} = \left(I_{res} + I_{cap}\right) = I_{res} + C_{t-s} \cdot \frac{\partial HV}{\partial t} \quad , \quad (6.1)$$

assuming a constant C_{t-s}. If a significant fraction of the total measured current is capacitive, the dose of electrons delivered to the resist may be much lower than desired. In particular, whenever the voltage changes abruptly, there can be a break in the exposure.

We measured a probe-sample capacitance as high as 2.4 pF. We estimate a contribution to the the total capacitance of less than 10^{-18} F from the tip itself (modeling the tip as a sphere after Sarid [8]). The cantilever, although quite close to the sample, is small in area (about 10^4 μm^2) and therefore contributes only a few fF. The major contribution to the tip-sample capacitance comes from the chip on which the cantilever is fabricated. The silicon chip is 3.6 × 1.6 mm, made large enough for ease of handling.

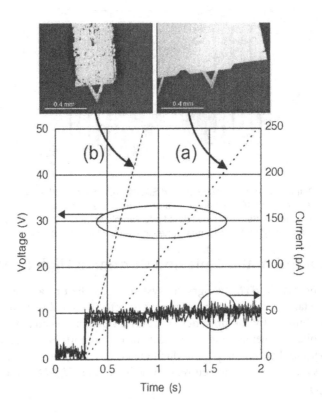

Figure 6.6: The effect of cantilever chip size on the voltage ramp during contact mode current-controlled lithography. The setpoint current was abruptly changed from 0 to 50 pA at time t=0.3 s. Curve (a) shows a voltage ramp of 30 V/s for the full chip (measuring 3.6 mm in width). The measured capacitance was 1.7 pF. Curve (b) corresponds to the chip shown above that was reduced to about 0.4 mm in width. The voltage here ramped at 91 V/s, indicating that the capacitance was reduced to 550 fF.

By reducing the size of this chip (on a grinding wheel), we lowered the total tip-sample capacitance to 550 fF. Figure 6.6 shows the effect of the chip size on the voltage ramp. We abruptly changed the setpoint current from 0 to 50 pA at time $t=0.3$ s. Although the difference between the setpoint and measured current (the error signal) was negligible, the measured current was purely capacitive during the voltage ramp. The speed of this ramp was limited by the tip-sample capacitance (for HV below the emission threshold, $I_{meas}=C_{t-s}\cdot\delta HV/\delta t$). The small chip was used for gate level lithography in the 100 nm pMOSFET fabrication (see Section 5.2) to ensure continuous patterning over 200 nm of topography (where the resist thickness changed by more than 60 nm and thus required a varying bias to maintain a fixed exposing current).

In order to further reduce the influence of the probe-sample capacitance, we built a circuit that compensates for the capacitive component of the current and ensures that the feedback responds primarily to the exposing current [Fig. 6.7(a)]. The compensation circuit generates a voltage (V_{comp}) proportional to the displacement current:

$$V_{comp} = R_{comp} \cdot C_{comp} \cdot \frac{\partial HV}{\partial t} \qquad (6.2)$$

V_{comp} is subtracted from the output of the current preamplifier (V_{meas}) to yield a voltage proportional to the current through the resist (V_{res}):

$$V_{res} = \left(V_{meas} - V_{comp}\right) = S \cdot I_{meas} - R_{comp} \cdot C_{comp} \cdot \frac{\partial HV}{\partial t} \qquad ,(6.3)$$

where S is the gain of the current preamplifier in units of V/A. The current control then feeds back on V_{res}, thus maintaining a constant current through the resist.

C_{comp} was chosen smaller than the probe-sample capacitance, although its precise value is not important since the compensation may be matched to the capacitance of a given system by adjusting R_{comp}. Figure 6.7(b) shows the effect of varying R_{comp} on the voltage ramp. Here the setpoint current was changed abruptly from 0 to 80 pA at $t=5$ s. In all cases the feedback immediately responded, making the measured current equal to the setpoint current.

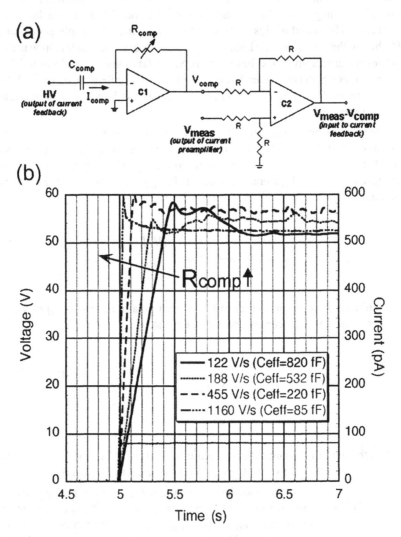

Figure 6.7: Capacitance compensation for high speed resist exposure.
(a) Diagram of the capacitance compensation circuit. (b) The effect of
capacitance compensation on the voltage ramp during lithography. As R_{comp} was
increased, the voltage bias more quickly reached that required for field emission.
The effective probe-sample capacitance was reduced from 820 fF   to 85 fF
using this compensation technique.

With no compensation (R_{comp}=0), the voltage increased slowly (122 V/s), corresponding to a tip-sample capacitance of 820 fF. For approximately 0.5 s there was no exposure. At a slow scan speed of 0.1 µm/s, this corresponds to a scan distance of only 50 nm; at a high scan speed of 1 mm/s, there would be no patterning until the tip traveled 0.5 mm, and this is clearly unacceptable. The voltage ramp can be dramatically increased by adjusting R_{comp} [Fig. 6.7(b)]. The fastest voltage ramp shown here is 1160 V/s, corresponding to an effective capacitance of 85 fF. The effective capacitance may be further minimized by fine-tuning R_{comp}, although it can never be completely eliminated and therefore continues to limit the response of the current feedback.

6.2.3 High Speed Lithography

The reduced effective capacitance allows the current feedback to keep the exposing current constant even at high scan speeds. We used this system to pattern Microposit SAL601 negative tone resist at scan speeds from 1 µm/s to 1 mm/s with various current setpoints. We etched the native oxide off silicon wafers, singed the samples, and primed the surface with vapor hexamethyldisilazane (HMDS) adhesion promoter prior to spin-coating the resist to a thickness of 65 nm. Details are given in Section 3.1.2. After exposure, the wafers were given a post exposure bake (PEB) for 1 min at 115 °C and developed in MF-322 for 10 min. The resist patterns were dry etched into the underlying silicon. Line width measurements were made from high-resolution scanning electron microscope (SEM) images of the etched features.

We found that the voltage required to emit a given current depends strongly on the scan speed. Figure 6.8(a) shows the tip-sample voltage bias necessary to achieve setpoint currents of 20 pA, 50 pA, and 100 pA at various scan speeds (solid lines). Constant force was maintained between the probe and the surface of the resist. At higher scan speeds, a higher voltage was required to maintain each current level. The tip-sample voltage dependence on scan speed (s) and setpoint current (I) can be approximated as:

$$V(I,s) = V_1(I) + 4\log(s) \qquad , \qquad (6.4)$$

where s has units of µm/s and $V_1(I)$ is the voltage bias necessary to achieve I at a scan speed of 1 µm/s. V_1 is approximately 39.4 V, 41.4 V, and 43.4 V for current setpoints of 20 pA, 50 pA, and 100 pA, respectively. Clearly the exposure changes the electrical properties of the resist. The exposure may induce electrical breakdown of the resist resulting in a lower impedance path between the tip and underlying sample. Therefore if the tip were scanned slowly (and thus had a significant dwell time over each pixel), the average voltage necessary to emit the desired current would be lower than if the tip were scanned quickly.

Figure 6.8(a) also displays the patterned line width for the different current and speed conditions (dashed lines). There is a maximum patterning speed corresponding to each exposing current. In Fig. 6.8(b) the same voltage and line width data are plotted versus exposure line dose (in units of charge per unit length, nC/cm). Notice the collapse of the line width data taken at different current setpoints. This indicates that the exposure dose is indeed the critical parameter for exposure. We observe optimum SAL601 exposure at line doses of 20–200 nC/cm (corresponding to line widths of approximately 30–120 nm). There is a practical upper dose limit of about 2000 nC/cm, above which the exposed patterns tend to "delaminate" from the substrate. We speculate that this delamination is due to stress in the resist film resulting from the high dose delivered.

Figure 6.9 shows how the current-voltage relationship depends on writing speed. Data for 10 µm/s (solid line) were acquired by measuring the applied bias necessary to achieve each current level while scanning with the tip in contact with a 65-nm-thick SAL601 resist film. We used Eq. (6.4) to generate the corresponding curves for speeds of 1, 100, and 1000 µm/s. Careful real-time adjustment of the applied voltage is necessary to generate the desired SPL current because of the strong dependence of emission on scan speed.

Figure 6.10 shows an SEM image of lines patterned in 65-nm-thick SAL601 resist at a scan speed of 0.5 mm/s and an emission current of 1 nA. The SEM image was taken after dry etch pattern transfer into the silicon substrate. The area shown is a section of a 10×10 µm line grating with 65-nm-wide lines on a 500 nm pitch. The entire grating was patterned in only 0.4 s. At conventional SPL speeds for local oxidation or resist exposure of 1–10 µm/s, the pattern would have taken several minutes to write.

6.2.4 Summary of High Speed SPL Using a Single Tip

We used electrons emitted from a scanning probe tip to pattern organic resist at high speeds. A compensation circuit was used to minimize the effect of the tip-sample capacitance that limited the effective bandwidth of the current feedback. This allowed us to demonstrate current-controlled lithography of SAL601 resist at speeds up to 1 mm/s. We have shown that the voltage required to generate the exposing current depends on the tip scan speed.

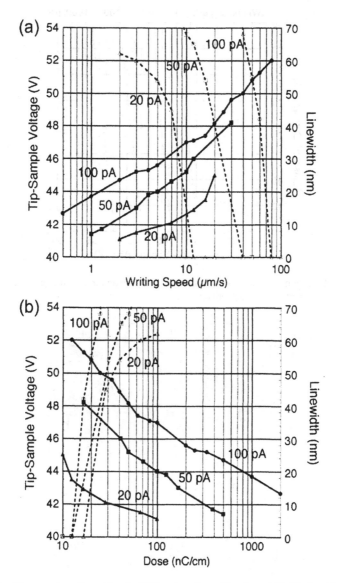

Figure 6.8: Tip-sample voltage bias (solid lines) and resulting line width (dashed lines) for different setpoint currents and scan speeds during contact mode SPL of 65-nm-thick SAL601 resist. (a) Data plotted versus scan speed. (b) Data plotted versus line dose.

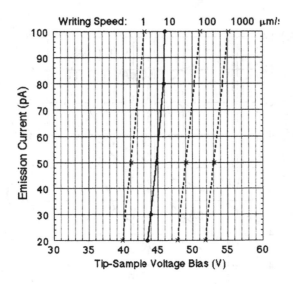

Figure 6.9: Emission current versus tip-sample voltage bias for 65-nm-thick SAL601 resist exposure at different writing speeds.

Figure 6.10: SEM micrograph of patterns written with a scanning probe in SAL601 resist at 0.5 mm/s and etched into the underlying silicon substrate. The lines are on a 500 nm pitch and were exposed with an emission current of 1 nA. The line width is approximately 65 nm.

6.3 References

[1]   P. Pai, W. G. Oldham, and C. H. Ting, "Process considerations for using spin-on glass," Proceedings of the Fourth International IEEE VLSI Multilevel Interconnection Conference (Cat. No.87CH2488-5), 364 (1987).

[2]   T. Nakano, K. Tokunaga, and T. Ohta, "Relationship between chemical structure and film properties of organic SOG," Extended Abstracts of PLANAR 94, Planarization Techniques for Submicron Technologies, Sunnyvale, CA, 6 June 1994.

[3]   I. Brodie and C. A. Spindt, "Vacuum microelectronics," Advances in Electronics and Electron Physics **83**, 1-106 (1992).

[4]   S. R. Manalis, S. C. Minne, A. Atalar, and C. F. Quate, "High-speed atomic force microscopy using an integrated actuator and optical lever detection," Rev. Sci. Instrum. **67**, 3294-3297 (1996).

[5]   S. C. Minne, S. R. Manalis, and C. F. Quate, "Parallel atomic force microscopy using cantilevers with integrated piezoresistive sensors and integrated piezoelectric actuators," Appl. Phys. Lett. **67**, 3918-3920 (1995).

[6]   S. C. Minne, G. Yaralioglu, S. R. Manalis, J. D. Adams, J. Zesch, A. Atalar, and C. F. Quate, "Automated parallel high-speed atomic force microscopy," Appl. Phys. Lett. **72**, 2340-2342 (1998).

[7]   S. C. Minne, Ph. Flueckiger, H. T. Soh, and C. F. Quate, "Atomic force microscope lithography using amorphous silicon as a resist and advanced in parallel operation," J. Vac. Sci. Technol. B **13**, 1380-1385 (1995).

[8]   D. Sarid, *Scanning Force Microscopy* (Oxford University Press, New York, 1991).

[9]   S. C. Minne, G. Yaralioglu, S. R. Manalis, J. D. Adams, J. Zesch, A. Atalar, and C. F. Quate, "Automated parallel high-speed atomic force microscopy," Appl. Phys. Lett. **72**, 2340-2342 (1998).

7   *On-Chip Lithography Control*

Our preferred method of scanning probe lithography (SPL) uses electrons field emitted from a micromachined probe tip in air to expose organic polymer resists. The pattern dimension is set by the electron exposure dose delivered to the resist. Control of the exposure dose has been achieved previously through external feedback circuitry. Typically the emission current is measured and compared with the desired (or setpoint) current. A signal is sent to adjust either the tip-sample voltage [1-3] or the tip-sample distance [4-6] in order to ensure that the measured current does not deviate significantly from the setpoint. In place of this feedback circuitry, we integrated a transistor current source onto the cantilever chip to control the electron exposure dose delivered to the resist. In this chapter we describe the design, fabrication, and operation of this integrated current source.

7.1 Background and Motivation

There are a number of advantages to integrating the current control onto the cantilever chip:

(1) Integrated current control simplifies and miniaturizes the SPL system, creating a small unit cell comprised of the electron emitter (tip) and exposure dose control (current source), eliminating the need for external circuitry.

(2) The bandwidth of the external control loop, which is hampered by parasitic capacitances, limits the lithography speed [7]. An integrated current source could allow higher speed patterning.

(3) Integration of the SPL control on-chip greatly facilitates the extension to parallel lithography using arrays of scanning probes. SPL throughput may be increased by patterning simultaneously with multiple probes.

We integrated a metal-oxide-semiconductor field-effect transistor (MOSFET) onto the cantilever chip to act as a current source for control of the emission current from the tip. In the saturation regime, the MOSFET drain-to-source current is independent of the drain-to-source bias and set by the gate voltage. This property allows us to use a MOSFET as a voltage-controlled current source. This approach is similar to that used to stabilize the emission currents in field emitter arrays operating in vacuum for display applications [9-11]. Researchers found that the emission current of transistor-controlled devices was more uniform across arrays and more stable over time than the current of voltage-controlled and resistance-controlled field emitters [11].

7.2 MOSFET Design Considerations

Traditionally during SPL a positive voltage is applied to the sample while the tip is grounded (often through a preamplifier that measures the current flow). The tip-to-sample bias generates a concentrated electric field at the tip. When this field is sufficiently large, electrons may be emitted from the tip through the process of field emission. The emission current is described by the Fowler-Nordheim equation [Eq. (3.1)] and depends strongly on the work function of the emitter and the electric field strength at the tip [12]. The electric field in turn depends on the tip shape, the voltage bias, and the tip-to-sample spacing. In air, the field emission current is inherently unstable as adsorbates may alter the local work function [13]. The active current feedback circuit is therefore necessary during SPL both to stabilize the field emission current and to set the appropriate current level for resist exposure. An integrated current source must also fulfill these requirements.

A schematic diagram of the integrated MOSFET current source is shown in Fig. 7.1. The MOSFET is built in the top silicon layer of a silicon-on-insulator (SOI) wafer and is a partially depleted n-channel enhancement mode device. The silicon tip and cantilever act as the drain of the MOSFET. The sample bias is set large enough to generate a sufficient electric field at the tip for field emission (Fig. 7.2). The transistor source terminal serves as the "source" of electrons for emission. The transistor regulates the current flow from the tip to the sample. With no bias to the MOSFET gate (gate voltage, V_g=0), the device is turned off and no electrons can flow out of the tip. When the transistor is turned on (for V_g > transistor threshold voltage, V_t), a conducting channel forms between the source and drain. The emission current is set by the transistor saturation current level.

The desired device properties include small saturation currents, a predictable and small threshold voltage, a high drain/body breakdown voltage, low subthreshold currents, small parasitic capacitances, fast switching speeds, and reliable device operation. Since the SOI device layer is thick, the channel is only partially depleted and the transistor behavior is similar to that of a conventional bulk device. The process simulator TSUPREM-4 [14] was used to determine the specific design parameters.

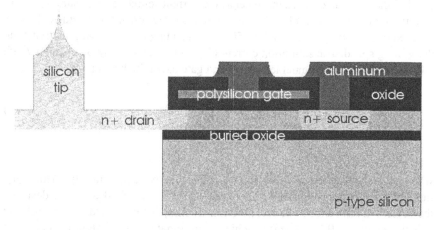

Figure 7.1: Schematic diagram of the cantilever with integrated transistor.

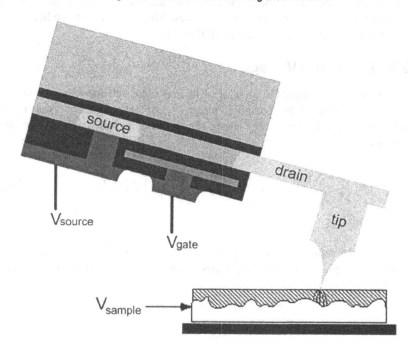

Figure 7.2: Diagram of the cantilever with integrated transistor used for SPL. The transistor regulates the electron dose delivered to the resist.

7.2.1 Saturation Current

For the integrated transistor to act as the current source for exposure dose control, the saturation current levels must be appropriate for SPL exposures. Typical SPL exposing currents are 10 pA to 50 nA, depending on the resist sensitivity and the writing speed. The saturation current (I_{dsat}) depends linearly on channel width (W) and inversely on channel length (L) and gate oxide thickness (t_{ox}):

$$I_{dsat} = \frac{\mu_n W \varepsilon_{ox}}{2 L t_{ox}} \left(V_g - V_t \right)^2 \qquad , \qquad (7.1)$$

where μ_n is the electron mobility and ε_{ox} is the gate oxide permittivity. [This equation holds under the conditions $V_g \geq V_t$ and $V_d > (V_g - V_t)$ for $V_s=0$.] We designed the devices to have a long channel (100 μm–10 mm), a narrow width (2–5 μm), and a thick gate oxide (0.5–1.0 μm) to achieve appropriately small saturation currents.[a] The transistors should also have a flat current versus voltage relationship in saturation so that they act as constant current sources. This idealized device behavior is most closely matched by long channel MOSFETs.

7.2.2 Threshold Voltage

A single voltage signal to the gate of the transistor sets the exposing current. We sought to minimize the threshold voltage (V_t) so that the gate bias during SPL could be a low voltage signal. For a long-channel FET, V_t is that of a MOS capacitor given by:

$$V_t = \Phi_{MS} + 2\left|\phi_{F_{sub}}\right| + \frac{t_{ox}}{\varepsilon_{ox}}\sqrt{4\varepsilon_{Si}qN_a\left|\phi_{F_{sub}}\right|} - Q_f\frac{t_{ox}}{\varepsilon_{ox}} \qquad , \qquad (7.2)$$

where ε_{Si} is the substrate silicon permittivity and Q_f is the oxide interface charge density. Φ_{MS} is the work function difference between a polysilicon gate and silicon substrate:

$$\Phi_{MS} = \phi_{F_{sub}} - \phi_{F_{gate}} \qquad , \qquad (7.3)$$

---

a. For example, a device with $L=1$ mm, $W=2$ μm, $t_{ox}=1$ μm, and substrate doping $N_A=3\times10^{15}$ cm^{-3} has $I_{dsat} \approx 4.5$ nA at $(V_g-V_t)=1$ V.

where ϕ_F is the Fermi energy:

$$\phi_{F_{sub}} = \frac{kT}{q}\ln\left(\frac{n_i}{N_A}\right) \quad \text{and} \quad \phi_{F_{gate}} = \frac{kT}{q}\ln\left(\frac{N_D}{n_i}\right) \tag{7.4}$$

for the p-type substrate and n-type gate, respectively. N_A is the substrate doping density and N_D is the polysilicon gate doping density [15]. V_t can be specified with appropriate choices of N_A and t_{ox}.

To minimize V_t while implementing an unusually thick gate oxide, we used a substrate with a low doping density and performed a threshold adjust implant in the transistor channel to ensure a positive but small V_t.[a] A forming gas anneal (FGA) was performed as a final step in the transistor fabrication process to lower Q_f. V_t variations among transistors should be minimized to enable uniform patterning from multiple tips.

7.2.3 Junction Breakdown

SPL generally employs organic resists of thickness 30-100 nm, requiring a tip-sample voltage bias of approximately 20-80 V for electron exposure. Thicker resist films are desirable to provide etch resistance during pattern transfer. The break-down voltage (V_{bd}) of the transistor drain-body junction determines the maximum usable resist thickness. Therefore we needed to maximize V_{bd} to avoid reverse-bias avalanche breakdown across the drain-body diode.

Breakdown occurs when the maximum electric field at the junction reaches a critical value (E_{crit}). The drain is heavily doped (n+) to achieve low resistivity and the body is p-type. For an ideal one-dimensional (1D) n+/p junction [16],[b]

$$V_{bd} \approx \frac{\varepsilon_{Si} E_{crit}^2}{2qN_A} \tag{7.5}$$

---

a. A device with t_{ox}=1 μm, N_A=3×10^{15} cm^{-3}, Q_f=q·N_f, and N_f=10^{11} cm^{-2} has V_t≈4.33 V. Because Q_f may vary with process conditions and could drive down V_t such that the FET becomes a depletion mode device (i.e., turned on for V_g=0), we performed a threshold adjust implant in the channel to raise N_A slightly and ensure V_t>0.
b. An ideal 1D n+/p junction with N_A=10^{15} cm^{-3} has E_{crit} ≈ 3×10^5 V/cm and V_{bd} ≈ 290 V.

For a two-dimensional junction, the maximum field depends strongly on the junction radius of curvature. We generated deep source/drain junctions by diffusing the implanted dopants completely through the top silicon layer in the SOI stack in order to reduce the maximum field at the junction.[a]

We also used a low substrate doping N_A to increase the junction depletion width and hence the junction breakdown voltage. The threshold adjust implant was performed throughout the channel *except* next to the drain. The depletion width, x_{dep}, of a n+/p junction is approximately [16]

$$x_{dep} = \left[\frac{2\varepsilon_{Si}}{qN_a} \left(V_{bi} + |V_A| \right) \right]^{\frac{1}{2}} \qquad (7.6)$$

where V_{bi} is the built-in voltage of the junction,

$$V_{bi} = \frac{kT}{q} \ln\left(\frac{N_A N_D}{n_i^2} \right) \qquad (7.7)$$

and V_A is the applied voltage.[b] Here N_A is the p-type substrate doping density and N_D is the n+ drain doping density. The threshold implant mask was designed such that the channel region at least one maximum depletion width from the drain did not receive the threshold adjust implant.

7.2.4 Off Current

We often make the simplifying approximation that no current flows between the MOSFET source and drain when the transistor is in its off state ($V_g < V_t$). However, there is in fact a finite subthreshold current flow (I_{d_subvt}) arising from inversion charge that exists at $V_g < V_t$. I_{d_subvt} depends exponentially on V_g [16] according to

$$I_{d\_subvt} = \mu_n \frac{kT}{q} \frac{W}{L} Q_n \left[1 - \exp\left(\frac{-V_g}{\xi\alpha} \right) \right] \qquad (7.8)$$

---

a. A diffused 3-μm-deep n+/p junction with $N_A = 10^{15}$ cm^{-3} has $V_{bd} \approx 80$ V.
b. A junction with $N_A = 10^{15}$ cm^{-3}, $N_D = 10^{20}$ cm^{-3}, and $|V_A| = 100$ V has $x_{dep} \approx 11.5$ μm.

---

assuming both the source and body are grounded. Q_n is the free charge density, which depends on ϕ_F and N_A. The transistor off current must be below the threshold for resist exposure. We would prefer that the magnitude of the subthreshold current be negligible compared with the exposing current, since it will provide a background electron exposure to the resist that reduces the image contrast. A small W/L ratio is useful in achieving this. The drain/body junction area should also be kept small to minimize the reverse bias leakage current across this diode.

7.2.5 Switching Speed

The maximum pixel rate for lithography (corresponding to the maximum writing speed) will be set by the switching speed of the integrated MOSFET. The transit time for electrons to travel from the source to the drain in an nMOSFET operating in the saturation regime is given by [16]:

$$T_{tr} = \frac{4}{3} \frac{L^2}{\mu_n \left(V_g - V_t \right)} \qquad . \qquad (7.9)$$

This simple formula predicts high switching times even for unusually long channel devices.[a] In reality the response speed of a FET is limited by the time needed to charge the capacitances associated with the transistor and elements to which it is connected.[b]

To achieve fast switching speeds we must minimize the drain-body junction capacitance, the gate-drain overlap capacitance, and the tip-to-sample load capacitance. The drain-body junction capacitance, C_j, is approximately [16]:

$$C_j = A \left[\frac{q \varepsilon_{Si} N_A}{2 \left(V_{bi} - V_A \right)} \right]^{\frac{1}{2}} \qquad (7.10)$$

and can be minimized by using a small junction area A and a low substrate doping N_A.[c] The gate-drain overlap capacitance, $C_{overlap}$, is given by [15]:

---

a. For a device with L=0.5 mm, μ_n=1500 cm^2/V-s, and $(V_g - V_t)$=5 V, the transit time would be 440 ns, corresponding to a switching speed of 2.3 MHz. A 2.3 MHz pixel rate and 50 nm pixel size corresponds to an tip writing speed of about 1.1 cm/s.

b. Using $I=C(\delta V/\delta t)$ we can estimate the minimum capacitance requirements to achieve a given switching speed. For SPL, we would like to achieve a pixel rate of 200 kHz, corresponding to a switching time Δt=5 μs. For ΔV=5 V and I=1 nA, the total capacitance must be below 1 fF to achieve a 200 kHz pixel rate.

$$C_{overlap} = \frac{\varepsilon_{ox} W x_j}{t_{ox}} \qquad , \quad (7.11)$$

where x_j is the source/drain junction depth.[a] The tip-to-sample capacitance includes a contribution from both the tip and the cantilever.[b] Use of tall tips increases the cantilever-to-sample distance, leading to a reduced capacitance. The combined effect of the parasitic capacitances limits the speed response of the integrated transistor.

7.3 Cantilever and Tip Design Parameters

The cantilevers were formed in <100> single-crystal silicon and doped n+ to act as the drain of the integrated transistor. We used a simple rectangular cantilever design of the geometry shown in Fig. 7.3. The end of the cantilever is pointed so that the corners do not accidentally touch the sample before the tip (as a result of a tilt error, for instance). The spring constant k of this cantilever is given by [17]:

$$k = \frac{E t^3 b}{2\left(2 L_1^3 + L_2^3\right)} \qquad , \quad (7.12)$$

where E is Young's modulus of the cantilever material,[c] and t is the cantilever thickness.

The tip-sample force is given by $F = k \cdot z$, where z is the vertical cantilever deflection. For SPL we may wish to operate the cantilever in the constant height mode (i.e., without real-time force feedback). Therefore we designed some compliant cantilevers so that the force exerted by the tip never becomes too substantial (such that the tip penetrates the resist film, for example).[d] Yet soft cantilevers, although exerting a smaller force, will not be able to track the sample surface at high speeds. Therefore we designed five different cantilevers with a range of spring

---

c. For $N_A = 10^{15}$ cm^{-3} and $V_A = 5$ V, the junction capacitance per unit area is ~ 4 nF/cm^2. For $A = 2.5 \times 10^{-5}$ cm^2, $C_j \sim 100$ fF. For $A = 10^{-6}$ cm^2, $C_j \sim 4$ fF. For $A = 4 \times 10^{-8}$ cm^2, $C_j \sim 0.2$ fF.

a. For a device with $W = 2$ μm, $t_{ox} = 1$ μm, and $x_j = 3$ μm, $C_{overlap} \approx 0.2$ fF.

b. We estimate a contribution to the total capacitance of less than 10^{-3} fF from the tip itself (modeling the tip as a sphere with radius 10 nm). The cantilever area is typically about 10^{-4} cm^2. For a 10 μm tall tip and a cantilever tilt angle of 15°, the cantilever-sample capacitance is roughly 2.5 fF.

c. $E \approx 1.7 \times 10^{11}$ N/m^2 for silicon where the cantilever is oriented along the <110> direction.

---

constants. All cantilevers have b=50 μm, L_2=25 μm, and $t \approx 1$ μm.[a] The cantilever lengths range from 50 – 400 μm with corresponding spring constants of 16 – 0.03 N/m (Table 7.1).

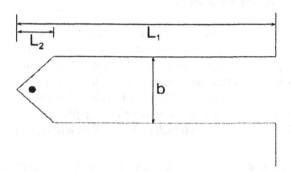

Figure 7.3:   Schematic diagram of the rectangular cantilever geometry.

Label	Cantilever Length, L_1	Spring Constant, k
A	200 μm	0.27 N/m
B	300 μm	0.08 N/m
C	100 μm	2.10 N/m
D	400 μm	0.03 N/m
E	50 μm	16.0 N/m

Table 7.1:   Five cantilever designs and corresponding spring constants for silicon cantilevers with b=50 μm, L_2=25 μm, and t=1 μm.

The integrated silicon tips must be tall (> 5 μm for low capacitance), sharp ($\leq$ 100 Å radius of curvature to achieve a small electron emission area), and have a high aspect ratio near the apex ($\leq$ 20° to ensure field concentration at the tip and for probing sharp surface features). The tip should be positioned near the end of the

d. For instance, if a sample has as much as 200 nm of topography (or, alternatively, of tilt relative to the scanner), and we want to maintain a tip-sample force below 50 nN, we must use a cantilever with a spring constant < 0.25 N/m.

a. The actual thickness may be different due to process variations. Even small variations in thickness can significantly affect the spring constant because k goes as t^3.

cantilever. We designed the chip on which the tip, cantilever, and transistor were fabricated such that it fits onto the scanning probe head of commercial instruments. The chip has a length of 3.5 mm and a width of 1.5 mm.

7.4 Fabrication Process

The cantilevers with integrated transistors were fabricated in the Stanford Nanofabrication Facility, a Class 100 clean room. The process is made up of four main tasks: (1) tip formation and cantilever definition, (2) front-end transistor fabrication, (3) back-end transistor fabrication, and (4) cantilever release. Steps (2) and (3) took advantage of conventional silicon device fabrication procedures and equipment. Steps (1) and (4) were micromachining processes that required unconventional steps such as deep trench etching and patterning on the back side of the wafer.

All patterning was performed with photolithography using either the Karl Suss MA-4 contact aligner or the Ultratech Model 1000 1:1 projection stepper. The full fabrication process required approximately 13 photomasks.

One challenge for the transistor fabrication was that the tall silicon tips, which were made first, had to be protected during allsubsequent process steps. This required the use of > 10-μm-thick photoresist films and/or protective oxide coatings on the tips.

Starting samples were 4-inch-diameter <100> silicon on insulator (SOI) wafers created using the separation by implantation of oxygen (SIMOX) process.[a] The top silicon was about 200 nm thick and the buried oxide (BOX) was about 370 nm thick. We grew 10 μm of boron-doped single crystal silicon in an epitaxial silicon reactor on top of the SOI stack.[b] The process flow is illustrated schematically in Fig. 7.4.

---

a. SIMOX SOI wafers were purchased from IBIS Technology Corporation, Danvers, MA.

b. Epitaxial silicon was grown by Moore Technologies, San Jose, CA. The epitaxial silicon was doped *in situ* with boron to yield a film with 15-20 Ω-cm resistivity (corresponding to a boron concentration of $1.5 \times 10^{15} - 3 \times 10^{15}$ cm^{-3}).

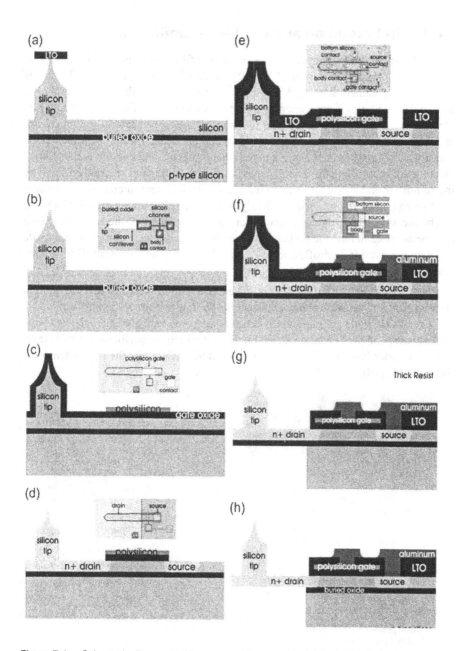

Figure 7.4:   Schematic diagrams of the process flow used to fabricate the devices.

7.4.1 Tip Formation and Cantilever Definition

A 1-μm-thick low temperature oxide (LTO) film was deposited by low pressure chemical vapor deposition (LPCVD) on top of the SOI stack. The LTO was patterned and etched, leaving 5-μm-diameter oxide circles that served as the tip etch masks. The tips were etched using a two-step process. First an isotropic SF_6 plasma etch was used to define the tip shape. This etched laterally about 2 μm under the LTO cap, leaving a 1-μm-wide silicon neck underneath. Next the silicon was anisotropically etched using a high-density plasma reactive ion etcher (RIE), creating the shaft of the tip. The thickness of the remaining top silicon layer will be the cantilever and transistor channel thicknesses (minus any silicon consumed during subsequent oxidation processes) [Fig. 7.4(a)]. The remaining silicon thickness can be measured using optical spectrophotometry; the length of the anisotropic etch may be adjusted to tune the tip height. After the tip etches were complete, the oxide cap was removed in hydrofluoric acid (HF). Figure 7.5(a) shows an SEM micrograph of the silicon tip after cap removal.

Nonuniformities in full-wafer plasma etchers tend to yield tips that are good in some regions of the wafer but not in others. To improve tip-to-tip uniformity, we left a relatively wide neck in the tip structure after the etch and sharpened the tips using a series of two oxidation steps, the first of which we performed at this stage. We grew about 0.4 μm of wet thermal oxide at 950 °C, below the reflow temperature of oxide [18].

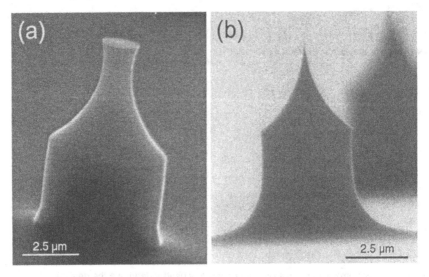

Figure 7.5: SEM micrographs of a silicon tip (a) after tip etch and oxide cap removal and (b) after oxidation sharpening.

The cantilevers were defined using photolithography with thick resist to protect the tall tips. The cantilever pattern was then etched into the thermal oxide. This oxide protected the tips during the next few steps.

7.4.2 Front-End Transistor Fabrication

We adopted a "mesa isolation" technique in which we etched away the top silicon everywhere around the device to electrically isolate each MOSFET [19]. We masked the silicon channel, source contact, and body contact with photoresist. Since the tips were protected with thermal oxide, we were able to use a thin photoresist layer here to achieve good lithography resolution (necessary to pattern a narrow transistor width). We next performed a silicon etch, stopping on the BOX [Fig. 7.4(b)]. During the etch, the patterned photoresist protected the channel, source contact, and body contact regions while the previously-patterned thermal oxide protected the tip and cantilever (which make up the transistor drain). The protective oxide was then removed.

Boron ions (dose=8×10^{11} cm^{-2}, energy=100 keV) were implanted throughout the nMOS channel (except next to the drain) to adjust the transistor threshold voltage. We defined the implant openings using photolithographic patterning of thick resist. The wafers were then annealed to drive the boron ions deep into the silicon channel (since the top layer of silicon is consumed during the subsequent oxidation step).

A 0.5- to 1.0-μm-thick film of wet thermal oxide was grown at 950 °C everywhere on the samples [Fig. 7.4(c)]. This step formed the transistor gate insulator and at the same time sharpened the tips. Next a 0.5-μm-thick blanket film of *in situ* doped polysilicon was deposited by LPCVD. We patterned and etched the polysilicon to define the transistor gate and gate contact regions [Fig. 7.4(c)]. The thick gate oxide was then selectively removed from the source and drain regions (including the tip). Figure 7.5(b) shows an SEM image of the silicon tip after oxidation sharpening. The tip height is approximately 8 μm, the cone angle is about 17°, and the tip radius of curvature is < 100 Å. Using this tip sharpening technique we were able to achieve good homogeneity of tips across the 4-inch-diameter wafers.

A phosphorous ion implant (dose=5×10^{15} cm^{-2}, energy=150 keV) was performed at both +7° and -7° tilts to dope the source and drain regions (including the tip) [Fig. 7.4(d)]. Another lithography step was used to open up contacts to the transistor body and the bottom silicon. We removed the oxide from these regions and implanted boron ions (dose=5×10^{15} cm^{-2}, energy=50 keV) to ensure ohmic metal-semiconductor contacts to the body and bottom silicon. A high-temperature furnace anneal was used to drive in the source/drain implants and to activate all dopants.

7.4.3 Back-End Transistor Fabrication

We passivated the devices with a 1 μm film of phosphorus-doped LTO [also known as phosphosilicate glass (PSG)] deposited by LPCVD. The PSG was annealed at 950 °C to reflow the glass, smoothing the transitions over sample topography. Contact holes were formed using photolithography of thick resist and anisotropic RIE of the PSG. We formed four contact holes per transistor: source, gate, body, and bottom silicon [Fig. 7.4(e)]. Contacts were made to the transistor drain only for some test devices. A contact and metal line from the drain are undesirable because they introduce additional parasitic capacitances that limit the speed response of the MOSFET. In SPL operation, the tip is held in close proximity to the resist surface; a voltage is applied to the sample and no drain contact is required.

A 1.5-μm-thick film of aluminum / 1% silicon was sputter deposited on the samples, filling the contact holes. The metal was patterned and wet etched to form the interconnects. The source and body of the transistor were tied together with one electrode that is grounded during device operation. We created three leads per device: source/body, gate, and bottom silicon [Fig. 7.4(f)], all of which terminate with 80 × 100 μm bond pads. The final transistor processing step was a forming gas anneal (FGA) performed at 400 °C. This reduces the gate oxide interface trap density and alloys the aluminum-silicon contacts.

We briefly discuss here two optional steps that may be performed prior to the FGA. One is a final passivation of the sample, in which another layer of LTO is deposited everywhere. Only the cantilevers and bond pads are opened using a wet pad etch. This LTO provides a protective overcoat to seal the devices from contamination and scratches. The second optional step coats the tips with metal. Although the tips are already highly doped, in some applications it might be desirable to have a conductive coating on the tips. A photolithography step opens regions in thick resist around the tips. The metal (typically titanium) is evaporated everywhere on the sample and is then lifted off, leaving metal only on the tips.

7.4.4 Cantilever Release

The cantilever release was performed using a bulk micromachining process. We protected the tip side of the wafer with a thick layer of photoresist. Then the backside of the wafer was patterned, and an STS Multiplex ICP deep trench etch was performed entirely through the 525 μm of bottom silicon. The BOX of the SOI wafer provided the etch stop [Fig. 7.4(g)]. The STS deep trench etcher uses alternating etching and passivation steps to achieve a highly anisotropic etch profile (Fig. 7.6). The backside etch removed the handle silicon beneath the cantilevers and divided the wafers into cantilever chips that can be manually separated and used in any scanning probe microscope.

After etching, the resist was stripped from both sides of the wafer in an oxygen plasma. Finally, the BOX was wet etched in HF, leaving free-standing cantilevers [Fig. 7.4(h)]. The wafers were carefully rinsed in deionized (DI) water and left to air dry. Figure 7.7(a) shows an SEM micrograph of two released cantilevers extending off of a chip with integrated transistors. The integrated silicon tip on a 1.5-μm-thick suspended cantilever is shown in Fig. 7.7(b). Here some oxide remains on the shaft of the tip, but the apex is clean.

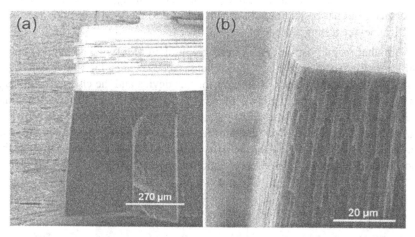

Figure 7.6:   SEM images of the cantilever chip. (a) The cantilevers were released using the STS etcher to etch anisotropically through the 525-μm-thick silicon handle wafer. (b) Higher magnification image of the vertical etch profile.

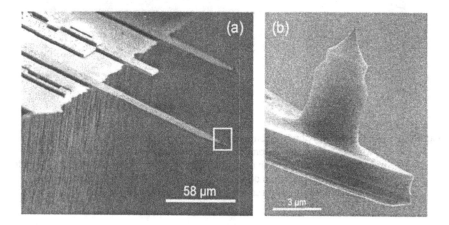

Figure 7.7:   (a) SEM micrograph of two released silicon cantilevers extending off a chip with integrated transistors. (b) Higher magnification image of the silicon tip on the released cantilever.

7.5 Device Characteristics

Figure 7.8 shows an optical micrograph of a completed cantilever chip. Two cantilevers extend off of each end of the chip. Each cantilever has a different length corresponding to a different spring constant, k, and acts as the drain of an integrated transistor, visible on the chip. These transistors have a gate length of 1 mm and a gate width of 2 μm. There is one "dummy" transistor between the two cantilevers on each end. This device has the same parameters (L, W, t_{ox}, etc.) as the neighboring devices, but has an additional contact and lead to the transistor drain to enable electrical device testing at a probe station.

A higher magnification optical micrograph of an nMOSFET integrated onto a cantilever chip is shown in Fig. 7.9(a). The distance between the source and drain, known as the channel length, is 100 μm for this device and the channel width is 5 μm. The integrated silicon tip is visible as a dot near the end of the cantilever and extends out of the image plane. SEM micrographs of the same device are shown in Fig. 7.9(b) and Fig. 7.9(c). The tilted images illuminate the topography on the sample. In Fig. 7.9(c) it is clear that the gate oxide and polysilicon wrap around the silicon channel, a result of the mesa isolation scheme adopted.

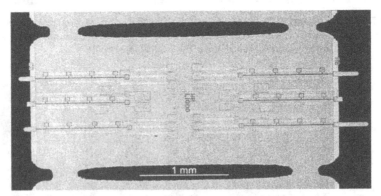

Figure 7.8: Optical micrograph of a completed 3.5 × 1.5 mm cantilever chip.
Two cantilevers extend from each end of the chip and are suspended. Each cantilever acts as the drain of an integrated transistor. The wafer was divided into chips during the backside deep trench silicon etch. The chips can be manually separated and used in a scanning probe microscope.

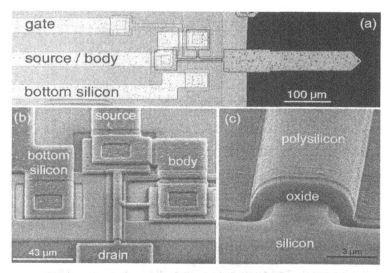

Figure 7.9: Images of a completed device. (a) Optical micrograph of a 100-μm-long nMOSFET integrated onto a cantilever chip. The three metal leads are labelled. (b) Tilted SEM image of the same device showing the sample topography. (c) Higher magnification SEM image showing that the polysilicon and gate oxide wrap around the silicon channel.

We formed linear channels (like those shown in Fig. 7.8 and Fig. 7.9) as long as 3 mm (on a 3.5-mm-long chip). Such long channel lengths were used to achieve small saturation currents (see Section 6.2.1 for details). The linear channel can be replaced by a "zig-zag" line to achieve a long length with minimized area. Figure 7.10 shows a 10-mm-long nMOSFET in which the channel meanders from the source to the drain. The meandering channel was defined in the top silicon of the SOI wafer and was entirely covered by the polysilicon gate. A reduced transistor area would be particularly important for dense arrays of probes. These devices behave just like linear channel transistors.

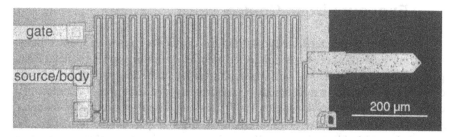

Figure 7.10: Optical micrograph of a 10-mm-long nMOSFET integrated onto a cantilever chip. The zig-zag gate design reduces the area that the transistor occupies.

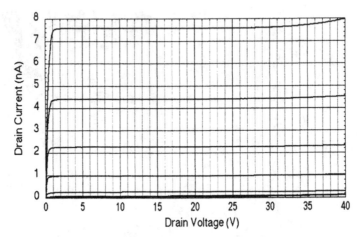

Figure 7.11: Electrical characteristics of an integrated long-channel nMOSFET as measured at a probe station. The gate voltage was varied from 0 to 3 V in 0.5 V increments. This device had L = 3 mm, W = 5 μm, and t_{ox} = 0.5 μm.

The integrated MOSFETs exhibited good electrical characteristics, with flat saturated currents, a threshold voltage of 0.5–2.5 V, and a breakdown voltage of > 45 V. Figure 7.11 shows the electrical characteristics of an integrated MOSFET with a gate length of 3 mm (linear channel), a gate width of 5 μm, and a gate oxide thickness of 0.5 μm. This data was acquired at an electrical probe station using a HP4155A parameter analyzer. The source/ body lead was grounded and the bottom silicon contact was left floating. This device has a threshold voltage of approximately 0.8 V. The saturation current is in the desired range for lithography with only a few volts applied to the gate. The drain current is independent of drain voltage until the junction breakdown at approximately 48 V. These device characteristics meet the requirements of an SPL current source.

7.6 Lithography with Integrated Transistor for Exposure Dose Control

For SPL with the integrated transistor, the cantilever chip was mounted on an insulating holder. Wire bonds connected electrodes on the holder to metal bond pads on the chip. The holder was mounted on a piezotube scanner with x, y, and z control. The transistor's source/body electrode was grounded through a commercial current preamplifier used to monitor the drain-to-source current during lithography. A negative bias may be applied to the bottom silicon to prevent the formation of a second channel (with the bottom silicon acting as a back gate), although we found that this was unnecessary. A positive voltage (typically 45-55 V) was applied to the film beneath the resist. The cantilever was lowered toward the resist-coated sample until it deflected slightly, corresponding to a tip-sample force of a few nN.

We scanned the tip over the resist surface. A positive bias was applied to the gate to turn on the transistor and set the current level. Figure 7.12 shows that the current measured during SPL of resist was identically the transistor saturation current. This transistor has a length of 1 mm, width of 5 µm, and gate oxide thickness of 1 µm. Data in Fig. 7.12(a) were taken at a probe station, while data in Fig. 7.12(b) were taken during patterning of 45-nm-thick PMMA resist. The good agreement shows that the transistor was indeed acting as a current source for control of the exposing current during lithography. The transistor off current was below the threshold for resist exposure; therefore a low voltage signal to the gate could be used to toggle the patterning on and off.

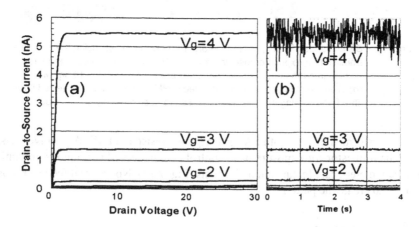

Figure 7.12: Drain-to-source current for an nMOSFET with L=1 mm, W=5 µm, and t_{ox}=1 µm (measured with the source grounded and the bottom silicon electrode floating). (a) Data taken at a probe station. (b) Data taken during lithography. During lithography, the tip was scanned in contact with a 45-nm-thick PMMA resist film. The current was measured at the source.

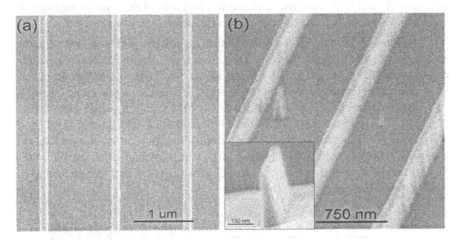

Figure 7.13: SEM images of lines patterned in organic polymer resists using the integrated transistor for exposure dose control. (a) SAL601 resist exposure and direct etching into the underlying silicon. Lines are approximately 110 nm wide. (b) Tilted SEM image of lines patterned in PMMA resist and transferred through lift-off and etching into the silicon. Lines are approximately 45 nm wide.

We performed SPL of both SAL601 negative resist and PMMA positive resist using the transistor for exposure dose control. We patterned continuous and uniform nanometer-scale features in both resists (Fig. 7.13). No external feedback was required to create the patterns.

7.7 Summary

We integrated a MOSFET onto a cantilever chip to act as a current source for electron field emission from the tip. The design parameters and fabrication procedure were presented. We demonstrated the use of this integrated transistor as the sole control electronics for stable SPL of nanometer-scale patterns in organic resist. This integrated current source eliminates the need for external feedback control circuitry and simplifies the lithography process. The integrated current source should facilitate the extension to parallel lithography with multiple scanning probes.

7.8 References

[1] K. Wilder, H. T. Soh, A. Atalar, and C. F. Quate, "Hybrid atomic force / scanning tunneling lithography," J. Vac. Sci. Technol. B **15**, 1811 (1997).

[2] H. Sugimura and N. Nakagiri, "AFM lithography in constant current mode," Nanotechnology **8**, A15 (1997).

[3] M. Ishibashi, S. Heike, H. Kajiyama, Y. Wada, and T. Hashizume, "Characteristics of scanning-probe lithography with a current-controlled exposure system," Appl. Phys. Lett. **72**, 1581 (1998).

[4] M. A. McCord and R. F. W. Pease, "Lift-off metallization using poly (methyl methacrylate) exposed with a scanning tunneling microscope," J. Vac. Sci. Technol. B **6**, 293-296 (1988).

[5] C. R. K. Marrian, E. A. Dobisz, and R. J. Colton, "Lithographic studies of an e-beam resist in a vacuum scanning tunneling microscope," J. Vac. Sci. Technol. A **8**, 3563-3569 (1990).

[6] K. Wilder, C. F. Quate, D. Adderton, R. Bernstein, and V. Elings, "Noncontact nanolithography using the atomic force microscope," Appl. Phys. Lett. **73**, 2527-2529 (1998).

[7] K. Wilder, H. T. Soh, A. Atalar, and C. F. Quate, "Nanometer-scale patterning and individual current-controlled lithography using multiple scanning probes," Rev. Sci. Instrum. **70**, 2822-2827 (1999).

[8] K. Yokoo, M. Arai, M. Mori, J. Bae, and S. Ono, "Active control of the emission current of field emitter arrays," J. Vac. Sci. Technol. B **13**, 491-493 (1995).

[9] G. Hashiguchi, H. Mimura, and H. Fujita,"Monolithic fabrication and electrical characteristics of polycrystalline silicon field emitters and thin film transistor," Jpn. J. Appl. Phys. **35**, L84-L86 (1996).

[10] J. Itoh, T. Hirano, and S. Kanemaru, "Ultrastable emission from a metal-oxide-semiconductor field-effect transistor-structured Si emitter tip," Appl. Phys. Lett. **69**, 1577-1578 (1996).

[11] S. Kanemaru, T. Hirano, H. Tanoue, and J. Itoh, "Control of emission currents from silicon field emitter arrays using a built-in MOSFET," Appl. Surf. Sci. **111**, 218-223 (1997).

[12] R. H. Fowler and N. Nordheim, "Electron emission in intense electric fields," Proc. R. Soc. London Ser. A **119**, 173-181 (1928).

[13]   B. R. Chalamala, Y. Wei, and B. E. Gnade, "The vital vacuum," IEEE Spectrum **35**, 50 (1998).

[14]   T-SUPREM4, Avant! Corporation, Fremont, CA.

[15]   S. Wolf, *Silicon Processing for the VLSI Era, Vol. 2* (Sunset Beach, California: Lattice Press, 1990).

[16]   R. S. Muller and T. I. Kamins, *Device Electronics for Integrated Circuits* (New York: John Wiley & Sons, 1986).

[17]   M. Torotonese, "Force sensors for scanning probe microscopy," Ph.D. Thesis, Stanford University (1993).

[18]   D.-B. Kao, J. P. McVittie, W. D. Nix, and K. C. Saraswat, "Two-dimensional thermal oxidation of silicon. II. Modeling stress effects in wet oxides," IEEE Trans. Elec. Dev. **ED-35**, 25-37 (1988).

[19]   J.-P. Colinge, *Silicon-on-Insulator Technology: Materials to VLSI* (Boston: Kluwer Academic Publishers, 1997).

[20]   T. Akiyama, U. Staufer, and N. F. de Rooij, "Wafer- and piece-wise Si tip transfer technologies for applications in scanning probe microscopy," IEEE J. MEMS **8**, 1 (1999).

[21]   D. Fletcher, K. Wilder, and C. F. Quate, "Post-processed epitaxial silicon scanning probe tips," unpublished.

[22]   P. Horowitz and W. Hill, *The Art of Electronics* (Cambridge: Cambridge University Press, 1995).

8   Scanning Probe Tips for SPL

This chapter presents the required features of probe tips for scanning probe lithography (SPL) and some novel methods for tip fabrication. For both field-induced oxidation and electron exposure SPL, probe tips must be electrically conductive and sharp to enable field concentration at the tip apex. For field emission of electrons from the tips, the workfunction of the tip material is significant. In the case of feedback control of the emitted current, the tips should be tall to ensure a small cantilever-to-sample capacitance. In this chapter we give details on different tip varieties: (1) "standard" silicon or metal-coated tips, (2) post-processed silicon tips, and (3) carbon nanotubes as scanning probe tips.

8.1   Silicon and Metal-Coated Tips

For most of the SPL experiments discussed in the early part of this book (Chapters 2-7), we used high-aspect-ratio doped-silicon tips for lithography.[a] Silicon tips are formed by wet or dry etching. Oxidation sharpening can be used to narrow the tip apex to a roughly 10 nm radius of curvature [1]. We coated some silicon tips with a thin film of metal (such as titanium, tantalum, or molybdenum), carbon, or diamond. Figure 8.1 is a transmission electron microscope (TEM) image of a doped-silicon tip coated with 20 nm of evaporated titanium (Ti), showing that the Ti covers well the full length of the tip. The radius of the silicon tip was about 10 nm. With the Ti coating, the tip radius was approximately 30-40 nm.

We experimented with boron-doped silicon tips and tips coated with metal, carbon, and diamond for field-emission of electrons and resist exposure. From Fowler-Nordheim theory (Chapter 3) we expect the emitted current to depend strongly on the emitter work function. We noticed a slight shift in the I-V curve as a function of tip material. Table 8.1 shows the average voltage necessary to emit 100 pA of current from different tips while scanning at 2 µm/s in contact with a 50-nm-thick film of PMMA. The titanium-coated tips required about 1 V more for emission than the highly-boron-doped silicon tips. Carbon-coated and diamond-coated tips required about 3-4 V less. The shifts in the voltage required to emit a given current do not correlate precisely with the emitter work function (Table 8.1). For example, we expected the titanium-coated tips to emit at a lower voltage than the

a. Ultralevers™ from Park Scientific Instruments (Sunnyvale, CA) or custom fabricated probe tips that were etched in silicon and oxidation sharpened.

doped silicon tips. The higher voltage necessary for emission from the titanium tips may reflect the increased tip diameter after the metal deposition. The emitted current signal from metal-coated tips appeared to exhibit less noise than that from the doped silicon tips and therefore was preferable for current feedback. Most of our SPL resist-exposure experiments were performed with titanium-coated tips

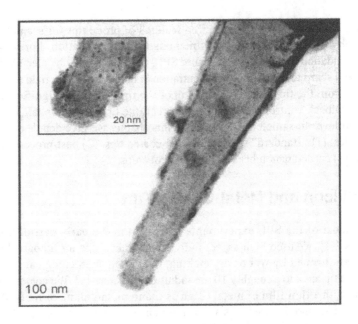

Figure 8.1:   TEM micrograph of a doped silicon probe tip coated with 20 nm of evaporated Ti. The 19 000 × magnification image shows the Ti covers the entire tip. The inset, at 120 000 × magnification, shows the Ti balls up slightly at the tip apex. The silicon tip radius was ~ 10 nm. With the Ti coating, the radius was 30-40 nm.

Tip Material or Coating	Voltage	Work Function
p+ silicon	45 V	5.17 eV
titanium[a]	46 V	4.3 eV
carbon[b]	42 V	2 – 3 eV
diamond[c]	41 V	3 – 5 eV

Table 8.1:   Average voltage necessary to emit 100 pA of current while scanning at 2 μm/s in contact with a 50-nm-thick film of PMMA.

a. Titanium (100 Å) evaporated on top of silicon.
b. Ion-beam deposited carbon (35-105 Å thick) on silicon [2][3].
c. Diamond deposited by unknown method on silicon by PSI [4][5].

8.2 Post-Processed Silicon Tips

The most significant fabrication challenge for the many cantilever structures with integrated components such as deflection sensors, current sources, and actuators is that the tip is created first from the thick top silicon of an SOI wafer. The tall silicon tip must then be protected during subsequent device processing steps using thick resist or other protective coatings. Thick resist processing is "nontraditional" and notoriously time-consuming. To facilitate the fabrication, it would be advantageous if the tip could be added *after* the device processing were complete.

To enable electron field emission from the tip apex, tips must be electrically conductive and sharp. In the case of cantilevers with integrated current sources (like the integrated transistor described in Chapter 7), the tip acts as the drain of the transistor so it must be electrically connected to the cantilever and transistor. Finally, the tips should be tall to ensure a small cantilever-to-sample capacitance. Several methods of "post-processing" tips have been previously demonstrated, but most require individual tip attachment and/or the use of an epoxy or glue to attach previously formed tips [6]. These methods are not suitable for batch fabrication of multiple devices with precise alignment requirements.

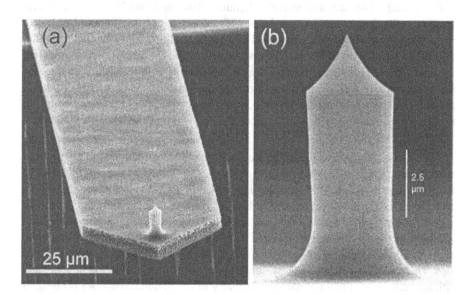

Figure 8.2: SEM micrographs of tips formed in epitaxial silicon. The process used to form these tips could be used to add tips after device fabrication. (a) Epitaxial tip on free-standing cantilever. (b) Tall, sharp epitaxial silicon tip.

We have shown that single-crystal silicon tips can be batch fabricated on wafers after significant device processing. After front-end transistor fabrication, the devices may be passivated with LTO. The LTO is patterned to open the cantilever regions. A thick layer of epitaxial silicon is grown from the top silicon layer in the SOI stack. The tips are then etched into the epitaxial silicon and oxidation sharpened [1][7]. Figure 8.2 shows images of a tip formed using this process. The epitaxial silicon growth is performed at 800–1100 °C and therefore must be completed prior to metal deposition. Nevertheless, this opens the possibility of using a commercial silicon foundry for device fabrication and then forming the tips and performing metal deposition and lithography as the final processing steps.

The use of a commercial silicon foundry would enable the implementation of more sophisticated current sources, such as a cascode or current mirror [8]. This should improve yield and repeatability, which is particularly important for arrays of probes.

8.3 Carbon Nanotubes as Scanning Probe Tips

Carbon nanotubes possess many properties that make them ideal candidates for use as scanning probe tips, including sharpness, high aspect ratio, high mechanical stiffness and resilience, and tunable chemical characteristics [9-14]. The atomic structure of carbon nanotubes can be considered a result of folding graphite layers into cylinders which may be composed of a single shell—a single-walled nanotube (SWNT)—or of many shells—a multi-walled nanotube (MWNT). A strong interest in the electrical properties of carbon nanotubes derives both from their unique status as atomically well-defined one-dimensional systems, allowing experimental studies of band structure, electron-electron interactions, electron-lattice coupling and electron localization, as well as from their potential as the basis of nanometer-scale electronic devices.

The first demonstrations of carbon nanotubes as tip material resulted from manual attachment of MWNTs [9] and ropes of SWNTs [10] to the pyramidal tip of a silicon cantilever for AFM imaging and lithography. In the early work, MWNTs were prepared by the direct current carbon arc method and manually bonded to the side of the tip using a soft acrylic adhesive. SEM micrographs of a probe and attached nanotube are shown in Fig. 8.3. In the results of these studies, nanotube tips exhibited convincing advantages over standard etched silicon or silicon nitride tips. They have proved to be highly durable and enable high-resolution imaging and lithography. Also, the long, narrow geometry of the nanotubes allows probing of high-aspect-ratio structures. Once a SWNT tip is realized, the diameter can be as small as 7 Angstroms.

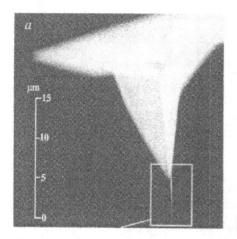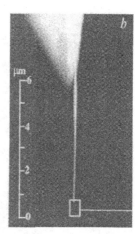

Figure 8.3: Single nanotube attached to the pyramidal tip of a Si AFM probe.
The image is reproduced with permission by H. Dai.

Despite the many attractive properties of nanotubes a tips, manual attachment is a time-intensive process. Furthermore, the reproducibility of the attachment process poses a serious problem. To overcome these problems, we describe here a fabrication strategy to integrate the growth of nanotubes into the cantilevers fabrication process.

8.3.1 Direct Synthesis on Silicon Pyramidal Tips

To integrate the nanotube growth directly onto silicon tips, we use a chemical vapor deposition (CVD) approach. This technique involves methane CVD synthesis of SWNTs [15][16] using a liquid-phase catalyst precursor material [17] consisting of three components: a mixture of inorganic chlorides, a tri-block copolymer serving as the structure-directing agent for the chlorides, and ethanol as the solvent.

The liquid phase catalyst precursor can be coated onto tips of AFM catilevers by dipping using a micro-pipette. After calcination and synthesis in methane, a well-oriented nanotube extending from the silicon pyramidal tip can be obtained. Typically, the SWNT are 1-20 μm in length beyond the tip apex. We shortened nanotubes to 30-100 nm (Fig. 8.4) in order to obtain rigid probe tips needed for AFM imaging. The shortening was achieved through arc-discharge under inert atomosphere.

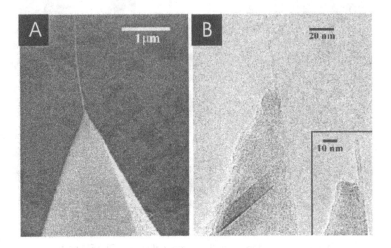

Figure 8.4: (a) SEM micrograph of SWNT after synthesis
(b) TEM images of oriented SWNT synthesized on a pyramidal tip before and
after shortening.

TEM imaging shows that SWNTs follow the surface of the silicon pyramids
before extending from the tip apex. We believe that Van der Waals interactions
between the nanotube and the pyramid surface is responsible for orienting the
SWNT. During the growth, SWNTs nucleate near the pyramid base and lengthen in
various directions. As growth terminates, the nanotubes adhere to the surface of the
pyramid and extend off the tip, thereby maximizing the tube-surface Van der Waals
interactions

We have investigated various growth and fabrication strategies to obtain a
large array of SWNT probe tips. One approach to synthesize properly-oriented
SWNTs on silicon tip arrays is to use contact printing [18]. In this technique, the
catalyst material is first spin-coated onto a flat polydimethylsilane (PDMS) stamp
as shown Fig. 8.5. Then the stamp is pressed against a silicon substrate containing
the silicon tip arrays to transfer the catalyst. Subsequently, the substrate is subjected
to calcination and methane CVD as described earlier.

The resulting SWNT probe array is shown in Fig. 8.6. The oriented SWNTs
were observed on roughly 30 percent of the silicon tips. The yield could be further
increased by optimizing the silicon tip strucure and improving the catalyst deposition process. The growth strategy may be extended to obtain SWNT probe tips on
arrays of AFM cantilevers leading to arrays of molecular probe tips at a large scale.

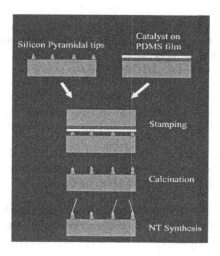

Figure 8.5: Contact printing process flow for fabricating arrays of SWNT probe tips.

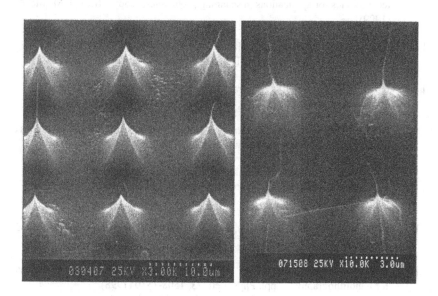

Figure 8.6: SEM micrographs of SWNTs on silicon pyramidal tips arrays fabricated by contact printing. A thin layer of gold (10 nm) was deposited on the sample to highlight the SWNTs.

8.4 References

[1]   D. Kao, J. McVittie, W. Nix, and K. Saraswat, "Two-dimensional thermal oxidation of silicon II. Modeling stress effects in wet oxides," IEEE Trans. Electron Devices **ED-34**, 1008 (1987).

[2]   G. A. J. Amarantunga and S. R. P. Silva, "Nitrogen containing hydrogenated amorphous carbon for thin-film field emission cathodes," Appl. Phys. Lett. **68**, 2529-2531 (1996).

[3]   R. D. Forest, A. P. Burden, S. R. P. Silva, L. K. Cheah, and X. Shi, "A study of electron field emission as a function of film thickness from amorphous carbon films," Appl. Phys. Lett. **73**, 3784-3786 (1998).

[4]   C. Nützenadel, O. M. Küttel, O. Gröning, and L. Schlapbach, "Electron field emission from diamond tips prepared by ion sputtering," Appl. Phys. Lett. **69** (18), 2662-2664 (1996).

[5]   W. A. Mackie and A. E. Bell, Proceedings of the 8th IVMC, Portland, OR (1995).

[6]   T. Akiyama, U. Staufer, and N. F. de Rooij, "Wafer- and piece-wise Si tip transfer technologies for applications in scanning probe microscopy," IEEE J. MEMS **8**, 1 (1999).

[7]   C. P. Ho, J. D. Plummer, S. E. Hansen, and R. W. Dutton, "VLSI process modeling-SUPREM III," IEEE Trans. Electron Devices **ED-30**, 1438 (1983).

[8]   P. Horowitz and W. Hill, *The Art of Electronics* (Cambridge: Cambridge University Press, 1995).

[9]   H. Dai, J.H. Hafner, A.G. Rinzler, D.T. Colbert, and R.E. Smalley, "Nanotubes as nanoprobes in scanning probe microscopy," Nature **384**, 147-150 (1996).

[10]   S. Wong, E. Joselevich, A. Wooley, C. Cheung, and C. Lieber, "Covalently functionalized nanotubes as nanometre-sized probes in chemistry and biology," Nature **394**, 52-55 (1998)

[11]   S. Wong, J. Harper, P. Lansbury, and C. M. Lieber, J. Am .Chem. Soc. **120**, 603-604 (1998).

[12]   H. Dai, N. Franklin, and J. Han, "Exploiting the properties of carbon nanotubes for nanolithography," Appl. Phys. Lett. **73**, 1508-1510 (1998).

[13]   J. Hafner, C. Cheung, and C. Lieber, Nature **398**, 761-762 (1998).

[14]   S. Wong, A. Wooley, E. Joselevich, and C. Lieber, "Functionalization of carbon nanotube AFM probes using tip-activated gases," Chem. Phys. Lett. **306**, 219-225 (1999).

References

[15]   J. Kong, A. M. Cassell, and H. Dai, "Chemical vapor deposition of methane for single-walled carbon nanotubes," Chem, Phys. Lett. **292**, 567-574 (1998).

[16]   J. Kong, H. T. Soh, A. Cassell, C. F. Quate, and H. Dai, "Synthesis of individual single-walled carbon nanotubes on patterned silicon wafers," Nature **395**, 878 (1998).

[17]   A. Cassell, *et al.* J. Am. Chem. Soc 121, 7975-7976 (1999).

[18]   Y. Xia and G. M. Whitesides, Angew. Chemie. (international edition) **37**, 551-575 (1998).

References

9   Scanning Probe Arrays for Lithography

In Chapter 6 we demonstrated dramatic improvements in the writing speed of a single tip, yet patterning throughput is still too low to make SPL a viable large-scale patterning technology. For example, a writing speed of 10 mm/s and a pixel size of 100 nm correspond to a pixel rate of 100 kHz (kilopixels per second). An exposure field measuring 1 cm × 1 cm contains 10^{10} pixels. If we raster scanned the tip over every pixel in the exposure field, it would take 10^5 seconds or about one day to cover the region. For comparison, today's deep ultraviolet (DUV) steppers pattern about 40 200-mm-diameter wafers per hour. Each wafer contains more than 200 1 cm × 1 cm exposure fields.

To further increase SPL throughput, we can pattern simultaneously with multiple probes. Minne *et al.* performed parallel oxidation lithography with an array of 50 cantilevers [1]. Since the electric-field-enhanced oxidation process is inherently slow, it may not be suitable for high-throughput patterning. Here we address the challenges encountered when the resist exposure scheme is extended to multiple tips.

9.1   Current-Controlled Lithography With Two Tips

In order to maintain the patterning reliability of our single tip resist exposure SPL system, we required individual control of the emission current from each tip. The current feedback system used for contact mode SPL draws the current to the preamplifier's virtual ground at the tip and applies a positive voltage to the sample (see Fig. 3.8). Alternatively, the current may be measured at the sample (at ground) while a negative bias is applied to the tip. In either case, the tip and sample are clearly coupled. Herein lies the challenge for multiple tip lithography.

In order to enable independent control of the emission current from multiple tips, we need a system capable of measuring the current at each tip *and* applying a high voltage to each tip. The sample, shared by all tips, must be held at a fixed voltage. Figure 9.1 depicts this scheme.

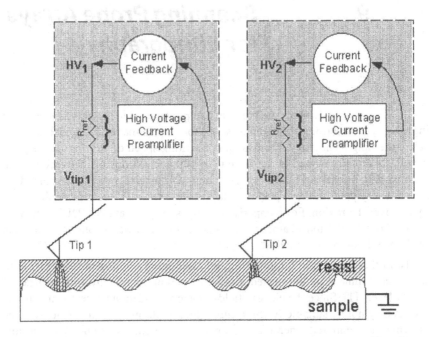

Figure 9.1: Schematic diagram of the current feedback scheme for multiple tip SPL. Since the sample is shared by the tips, it must be held at a fixed voltage such as ground. The current is measured at each tip and a high voltagebias is also applied at each tip.

9.1.1 High-Voltage Current Preamplifier

A high-voltage current preamplifier determines the current flow from tip to sample by precisely measuring the voltage drop across a large reference resistor. The circuit design, based on the instrumentation amplifier configuration, was chosen because of its high input impedance and its high common mode rejection ratio (CMRR) [2].

We designed and built a two-channel current preamplifier capable of low-noise current measurements at high voltages. Figure 9.2 shows a circuit diagram of one channel. All operational amplifier stages are high voltage devices. Device U2 in particular must have an extremely high input impedance and low bias currents since its noninverting input is in parallel with the tip/resist system. We used the Apex Microtechnology high power FET input amplifier PA85 for U1 and U2, and Apex's PA87A for U3, U4, and U5.

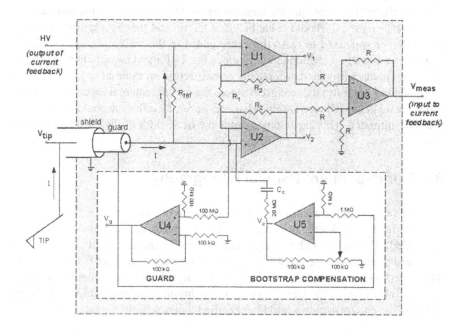

Figure 9.2:   Circuit diagram for the high-voltage current preamplifier.

The circuit gain is controlled by the ratio R_1/R_2 and can be varied without affecting the input impedance or the circuit CMRR:

$$V_{meas} = (V_1 - V_2) = \left(1 + \frac{2R_2}{R_1}\right)(V_{tip} - HV) = \left(1 + \frac{2R_2}{R_1}\right)R_{ref}I \qquad (9.1)$$

We chose the sensitivity of the high voltage preamplifier to be 10^9 V/A. The minimum detectable current is set by the size of the reference resistor (R_{ref}). We selected a 100 MΩ, 1% tolerant precision resistor for R_{ref}. The preamplifier therefore provides accurate measurements from 50 pA to 10 nA, which spans the appropriate current range for SPL. The high voltage preamplifier has a CMRR of almost 90 dB.

The cable capacitance and the input capacitance of U2 are in parallel with the tip-resist system, compounding our previous capacitance problem. We could compensate for these as described in Section 5.1.2. However, since the magnitude of the additional capacitances far exceeds the original tip-sample capacitance, we cannot expect to reduce the effective capacitance to a tolerable level in this way. Here we show an alternative technique for minimizing the effect of these capacitances (Fig. 9.2). First, we used a triaxial cable between the tip and the preamplifier, where a guard driver was used to raise the cable guard to the tip voltage. Second, we employed a bootstrap technique to minimize the 4 pF input capacitance of U2 [2]. A potentiometer was varied to force the current across an external capacitor (C_c) to be equal (and opposite) to the current across the op-amp internal input capacitance. Since the input capacitance is somewhat non-linear, its influence could not be completely eliminated. This capacitance limits the bandwidth of the current feedback system.

9.1.2 Independent Parallel Lithography

The two-channel current preamplifier was used in conjunction with two identical analog feedback circuits (see Fig. 2.9) to perform independent current-controlled lithography with two tips. The integral feedback circuit compared the measured current to the setpoint current and varied HV to minimize the error signal. The current setpoint was enabled with a voltage signal from the computer that controls the movement of the probe in order to synchronize the lithography with the tip motion. The current setpoint could be specified independently for each tip.

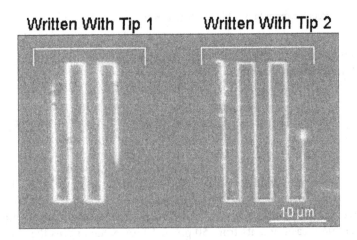

Figure 9.3: Developed SAL601 resist lines patterned by two tips simultaneously. The exposing current from each tip was independently controlled.

The cantilevers used were 2×1 arrays of micromachined silicon tips fabricated by S. C. Minne [3]. It is important that the two cantilevers are electrically isolated so that the high voltage applied to each tip does not cause significant current flow between the tips. The tips were scanned as a unit along the designated path in the constant height AFM mode. A result of parallel lithography with two tips is shown in Fig. 9.3, where the pattern on the left was written with tip 1 while the pattern on the right was written simultaneously with tip 2. This current control scheme may be extended to additional tips operating in parallel.

9.1.3 Summary of Progress on Parallel Lithography

We reported progress toward high-throughput nanolithography using scanning probes. We extended the current feedback scheme to two tips, where the exposing current from each tip was independently controlled. Multiple tip control required a new current preamplifier design with internal capacitance compensation in order to measure currents at high voltages. The independent current feedback allows different setpoint currents to be applied to each tip for individual dose and/or line width control. We demonstrated parallel, current-controlled SPL of SAL601 resist with two tips.

9.2   Massively Parallel Arrays for Lithography

We envision an SPL system that scans a massively parallel array of probes over a sample as illustrated in Fig. 9.4. The large array would be scanned as a unit in the plane of the sample, but each tip would be individually addressable for lithography. Both one-dimensional [4][5] and two-dimensional arrays [6] of scanning probes have been micromachined and used for parallel imaging, lithography, and data storage applications. Reliable and controlled resist exposure with each probe in a large array has not yet been demonstrated. The individual current feedback presented in Section 6.2 may provide one solution, although the task of replicating the external circuitry for hundreds or thousands of probes is daunting. Here we consider briefly how large an array would be required for SPL to provide reasonable throughput for next generation lithography (NGL).

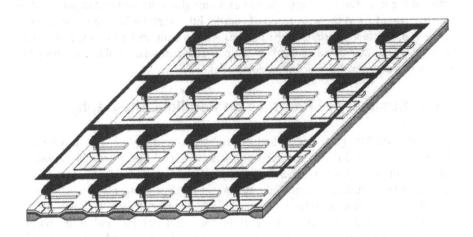

Figure 9.4: Schematic diagram of a two-dimensional array of probes patterning a surface.

9.2.1 Exposure Time for Different Size Arrays

The SPL exposure time depends on the pixel size (d_{pixel}), the pixel rate (R_{pixel}, in pixels/sec or Hertz), and the density of tips (D_{tips}) operated in parallel. The number of pixels per exposure field, N_{pixel}, is simply the exposure field area (A_{field}) divided by the pixel area (A_{pixel}):

$$N_{pixel} = \frac{A_{field}}{A_{pixel}} \qquad , \quad (9.2)$$

where $A_{pixel} = d^2_{pixel}$. The pixel rate is related to the linear scan speed of the tip (S) by

$$R_{pixel} = \frac{S}{d_{pixel}} \qquad , \quad (9.3)$$

N_{tip} is the number of tips per exposure field. The tip density, D_{tip}, is

$$D_{tips} = \frac{N_{tips}}{A_{field}} \qquad\qquad . \quad (9.4)$$

Therefore the exposure time per field (in seconds) is given by

$$t_{field} = \frac{N_{pixel}}{R_{pixel} N_{tips}} = \frac{1}{R_{pixel} D_{tips} A_{pixel}} = \frac{1}{D_{tips} d_{pixel} S} \qquad . \quad (9.5)$$

Figure 9.5 shows the time to pattern a 1 cm^2 area as a function of the pixel rate (or scan speed, above) for different tip densities. Here we assumed a pixel size of 100 nm. It can be seen that even with all of the tips scanning at 10 mm/s, we need a high tip density of 1000 tips/cm^2 to pattern the region in a manageable period (100 s). This patterning time only increases when the pixel size is further reduced.

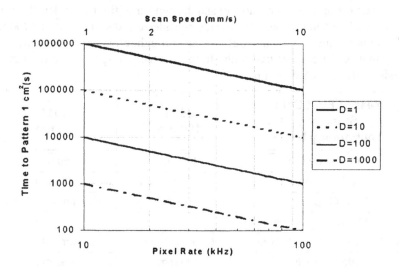

Figure 9.5: SPL exposure time to pattern a 1 cm^2 area using 100 nm pixels. Data shown for tip scan speeds of 1–10 mm/s, corresponding to pixel rates of 10–100 kHz. We show the exposure time for probe densities from 1 tip/cm^2 (top curve) to 1000 tips/cm^2 (bottom curve).

9.2.2 SPL Throughput Using Cantilever Arrays

We calculated the SPL throughput assuming that an array of tips covers a single exposure field. The array would be used to pattern one field, then the wafer would be stepped to shift the next field under the tip array. Patterning would continue until all N_{field} fields on the wafer were exposed, an approach similar to that used in conventional optical steppers. The raw throughput, T_{raw}, of such a system is given by

$$T_{raw} = \frac{3600}{t_{field} N_{field}}$$

(9.6)

in units of wafers per hour. (Here we neglect overhead time, stepping time, etc.)

As an example, we consider the case where the tip array covers an exposure field equal to the microprocessor chip size for the 100 nm technology node as displayed in Table 9.1. Each chip measures 20.5 × 20.5 mm. About 140 such chips fit on each 300-mm-diameter wafer. Fig. 9.6 shows the raw throughput as a function of pixel rate (or tip scan speed, above) and the number of tips (or tip density) using 100 nm pixels. Even the most dense array considered (1000 tips/cm^2 or 4200 tips total) cannot achieve even a 1 wafer/ hour throughput. Clearly the achievable throughput is not competitive with alternative technologies if the tip array is extended only over a single chip.

Year of First Chip Shipment Technology Node	2005 100 nm	2011 50 nm
Dense Line CD	100 nm	50 nm
MPU[a] Chip Area[b]	420 mm^2	600 mm^2
MPU Chip Size	20.5 mm x 20.5 mm	24.5 mm x 24.5 mm
Optical Exposure Field Area	1000 mm^2	1100 mm^2
Optical Exposure Field Size	25 mm x 40 mm	25 mm x 44 mm
Wafer Diameter	300 mm	450 mm
# MPU Chips Per Wafer	~ 140	~ 220

Table 9.1:   SIA International Technology Roadmap for Semiconductors projections for future critical level lithography requirements [7].

a.MPU=microprocessor unit.
b.Year 2 chip size.

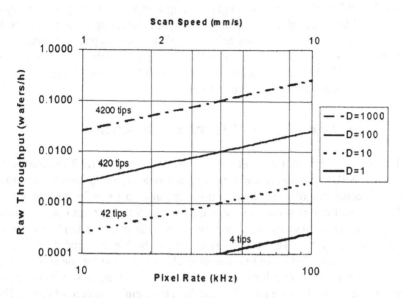

Figure 9.6: SPL throughput versus pixel rate (or tip scan speed) for chip-scale tip arrays. The array covered the 420 mm² chip. The pixel size was 100 nm.

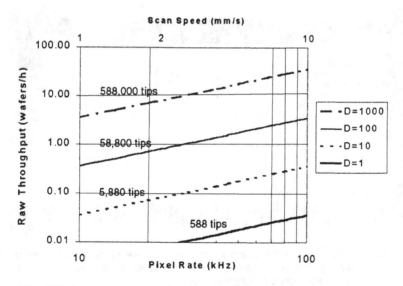

Figure 9.7: SPL throughput versus pixel rate (or tip scan speed) for wafer-scale tip arrays. The array covered the 300 mm wafer. The pixel size was 100 nm.

If we extend the tip array over the entire wafer, the situation improves. Now the exposure field area is equal to the entire patterned area on the wafer and

N_{field}=1. Fig. 9.7 shows the SPL throughput using a wafer-scale array. Large, dense arrays can achieve throughput competitive with other lithography technologies. In order to use such massive cantilever arrays for reliable lithography, we need a simple method for independently controlling the exposure by each probe. Ideally, the technology should be easily extendable from a small array to a large array. We present a possible solution in the next chapter.

9.3   Integrated Current Control for Arrays

The on-chip transistor current source described in Chapter 7 facilitates the extension of SPL to parallel lithography with multiple scanning probes since each "unit cell" (consisting of a cantilever and single transistor) may be replicated with standard semiconductor processing techniques. Fig. 9.8 shows an optical micrograph of an array of 10 cantilevers, each in series with an integrated transistor for exposure current control. The cantilevers are spaced by 100 μm, and the transistor gate length is 3 mm. There are two common electrodes: one for the source/body (usually grounded) and one for the bottom silicon. Each transistor has a single active lead for controlling the gate bias (and hence the emission current level). This allows independent control of the exposure dose from each tip in the array.

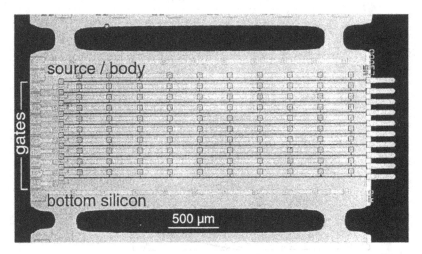

Figure 9.8:   Optical micrograph of an array of 10 cantilevers, each in series with an integrated transistor for exposure current control. The cantilevers are spaced by 100 μm, and the transistor gate length is 3 mm.

9.4   Two Dimensional Arrays: Process Development

To achieve the high probe densities discussed in Section 8.2, we must move to two-dimensional (2D) arrays of cantilevers. The fabrication and operation of a 2D array of cantilevers pose unique problems which are not encountered in single cantilevers or in a 1D array. 2D array design and fabrication are addressed in this section.

9.4.1 Enabling Technologies

A typical operating set-up of a 1D array is depicted in Fig. 9.9(a). The probe array is brought to the sample at an angle (usually 15°) which creates the separation between the sample and the chip containing the cantilever arrays. In this configuration, the contact pads can be fabricated on the same side as the silicon probe tip; the separation created by the approach angle permits wirebonding from the electrode pads to the external circuitry.

A schematic diagram of a 2D array set-up is shown in Fig. 9.9(b). In this configuration, the chip containing the probe arrays lies on a parallel plane to the surface of the sample, and the only separation between the probe array and the sample are provided by the height of the tip. Thus the first necessary component of a 2D array is the fabrication of a tall (>10 μm) probe tip with a high aspect ratio. The tall tips were formed using a series of isotropic and anisotropic etching steps followed by oxidation sharpening.

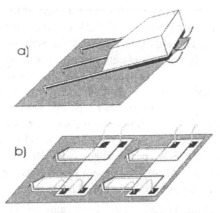

Figure 9.9:   Schematic diagram of array operation. (a) 1D array of probes and (b) 2D array of probes.

The second challenge in the fabrication of 2D array of AFM probes stems from the need to release an array with a high density of cantilevers. When fabricating a single probe or a 1D array of probes, anisotropic wet etching has been extensively used to release the cantilevers near the end of the fabrication process. The crystal direction dependent silicon wet etchants such as Sodium Hydroxide (KOH), Ethylene Diamene Protocatacol (EDP), and Tetramethyl Ammonium Hydroxide (TMAH) that were typically used creates a 54.7° angle to the surface of a (100) silicon, because of the etch rate difference between the (100) planes and (111) planes [8][9]. The 54.7° angle does not degrade the probe density in a 1 D array, however, in 2D arrays this angle occupies and unacceptably large area. Therefore, a vertical sidewall anisotropic etching is required to increase the probe densities.

The third challenge stems from the fact that the interconnects and contact pads must be placed on the opposite side of the wafer from the scanning tip. Unlike the 1D array, the separation between the probes and the sample is not sufficiently large to wirebond directly to the contact pads. A robust method for electrically connecting the front side of the wafer to the backside is necessary to operate a 2D array.

In this section, we describe the through-wafer via (TWV) technology together with the probe release by anisotropic reactive ion etching (RIE). The TWV technology can be used for a wide range of applications including interconnects in circuits, three dimensional (3D) packaging of chips, and 3D electrical and MEMS structures.

9.4.2 Anisotropic Through Wafer Etching

The deep silicon trench etch in this work was performed with a commercially available, high density low pressure (HDLP) reactive ion etching (RIE) system from STS Limited, UK. This etcher uses separate RF sources for the plasma generation (coil) and ion acceleration (substrate platen). It is capable of high anisotropy which is achieved by alternating between etching and passivating processes. During the etch cycle, it applies 600 W to the coil, and 120 W to the platen with an SF_6 flow rate of 130 sccm at 15 mTorr chamber pressure. During the subsequent passivation cycle, 85 sccm of C_4F_8 is flowed at the same plasma power and pressure with no platen power. The two cycles are repeated with 11 seconds of etching and 8 seconds of passivation to achieve near vertical walls [10]. The etch rates are dependent on the feature size and density; for 30 μm square vias the etch rate is about 2.2 μm/min and for 100 μm by 500 μm trenches, it is 4.5 μm/min.

This system of etching is used for both dry cantilever releasing and TWV formation. The performance of the etcher is demonstrated by etching 30 μm/side square vias through a 525-μm thick wafer, which translates to an aspect ratio of greater than 17:1. Optical micrographs of both sides of the wafer after RIE are shown in Fig. 4.2.

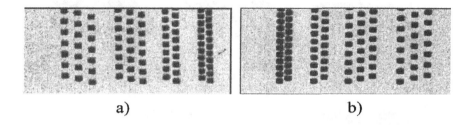

a)   b)

Figure 9.10:   High aspect ratio deep silicon trench etching.
Optical micrographs of 30-µm-wide trench arrays with different pitch:
(a) front, and (b) back sides of the wafer. The wafer is 525 µm thick.

9.4.3 Through-Wafer Via Process

There have been previous attempts to microfabricate through wafer vias (TWV's) [11],[12] using anisotropic wet etching. However, the density of TWV's were limited by the slopes created by the etchants. In this work, we describe TWV's with high wiring densities fabricated with reactive ion etching of the silicon substrate.

The challenge for the fabrication of high density TWVs originate from the fact that the topographies and aspect ratios are quite extreme. Besides the anisotropic, high-aspect ratio deep trench etching, one needs to electrically isolate each via, metallize, and perform resist processing on both sides of the wafer in the presence of deep (> 500 µm) and high-aspect-ratio (>17:1) trenches. Many conventional planar processing techniques are not suitable at such extreme geometries, and unconventional fabrication techniques such as electro-deposited resist processing were necessary. In this section a detailed process flow for high wiring density TWVs is described.

The substrate was a 4-inch diameter, 525-µm thick, 10 Ω-cm, p-type, double-sided polished silicon wafer. First, we spun on a 16-µm-thick photoresist film (AZ 4620, AZ photoresist products, Sommerville, NJ, USA), and patterned 30 µm/side squares using a contact mask aligner. The exposure was done in 8 steps of 10 s exposures with a 20 s delay between the exposure steps to prevent nitrogen bubbling in the thick resist. The resist was then developed in a 4:1 mixture of de-ionized water and AZ 400K developer for 6 min at room temperature. The post-develop bake was performed at 110 °C in a convection oven for 60 min.

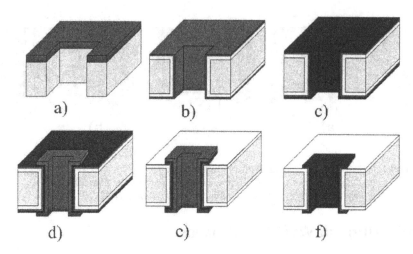

Figure 9.11: Through-wafer via process flow. (a) TWV is dry etched, (b) oxide isolation and polysilicon sticking layer are deposited, (c) Cu is deposited by CVD and electroplating, (d) photoresist is electroplated and patterned on both sides, (e) copper and polysilicon are etched, and resist is stripped.

The through-wafer deep silicon trench etch was performed with the HDLP RIE system described in the previous section with the thick photoresist as the mask. We etched square vias with 30 µm/side through the 525-µm-thick wafer. The resulting structure is schematized in Fig. 9.11(a)

After stripping of the thick photoresist, electrical isolation in the via was achieved by growing a 1 µm thick thermal oxide. We next deposited 1.5 µm of undoped polysilicon by low pressure chemical vapor deposition (LPCVD) in a furnace at 620 °C [Fig. 9.11(b)]. The oxide growth and LPCVD polysilicon deposition produced conformal films, which means that the thickness of each layer was the same on both sides of the wafer as well as on the sidewalls of the etched vias. Next, a 250-nm thick CVD copper was conformally deposited. Since the adhesion between thermal oxide and copper was poor, we required a polysilicon sticking layer between the CVD copper and the oxide. Cross-sectional SEM micrographs of the TWV are shown in Fig. 9.12.

Following the CVD copper as the seed layer, a 6-µm thick copper film was electroplated using a laboratory-built conformal copper electroplating system [Fig. 9.11(c)]. The resulting sheet resistance was 2.8 mΩ/sq.

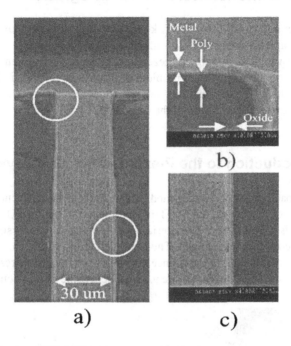

Figure 9.12:   Cross sectional SEM micrographs of a TWV.
The metallization is 1.5 µm in this sample. (a) The via is 30 µm wide and 525 µm
deep. Deposited films have good step coverage (b) at the corners and (c) on the
sidewalls.

A conventional spin-on type photoresist cannot be used for patterning both sides of the wafer. First, it causes streaking that leads to problems during exposure. Second, spin-on resist films cannot conformally coat the sidewalls of the vias, and this leads to inadequate masking of the copper metallization inside the vias. To overcome these problems, we used an electroplating technique. We electro-deposited a commercially available photoresist (PEPR 2400, Shipley Co, MA). The temperature of the electroplating bath was controlled at 32 °C, the spacing between the plating electrodes was 10 cm, and 300 V was applied for 20 s to achieve 7.2 µm thickness. The resist was exposed on a standard mask aligner and developed in a potassium carbonate (7.6 g) / de-ionized water (1 liter) solution at 35 °C [Fig. 9.11(d)]. Using the electro-deposited resist as a mask, the copper layers were wet etched [Fig. 9.11(e)] and then the polysilicon was plasma etched using SF_6 chemistry [Fig. 9.11(f)].

9.5 Two Dimensional Arrays: Integration

In this section, we describe the work on fabricating a 2D array of AFM probes. To make large area scanning feasible without the need for aligning external sensors, the probes must include integrated piezoresistors for deflection sensing. Furthermore, to alleviate the geometrical problem of electrically connecting the sensors, we integrated the TWV's so that the contacts are accessible from the backside of the wafer. We describe the design of the devices and report on their fabrication and performance.

9.5.1 Introduction to the Piezoresistive Cantilever

The approach of using an integrated piezoresistor as the deflection sensor in a silicon cantilever was pioneered by Marco Tortonese at Stanford University [13]. Since silicon is a piezoresistive material, when the cantilever is stressed by the deflection, the resistance changes. Thus, the deflection of the cantilever can be determined by measuring the resistance change in the integrated resistor through the use of a Wheatstone bridge. Figure 9.13 is a schematic of the set up.

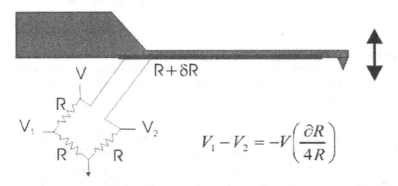

Figure 9.13: Schematic of the piezoresistive cantilever.

The piezoresistive cantilever can achieve sub-Angstrom resolution in deflection and thus is sufficient for atomic resolution imaging. It is our goal in this project to integrate the piezoresistive deflection sensors with the TWV's and fabricate a 2D array of AFM probes.

9.5.2 Design and Modeling

The piezoresistive effect can be expressed as:

$$(9.7)$$

$$\frac{\Delta\rho}{\rho} = \pi \frac{P(x-L)y}{I} \qquad ,$$

where ρ is the resistivity, π is the piezoresistive coefficient, P is the load at the tip, L is the length of the cantilever, I is the moment of inertia of the cross section, and y is the distance from the neutral axis. A typical piezoresistive cantilever design is provided in Fig. 9.14.

A detailed analysis of the piezoresistive cantilever performance including the spring constant, fractional change in resistance per unit displacement, and the minimum detectable deflection with such a geometry was performed by Tortonese [14], and a summary is provided here.

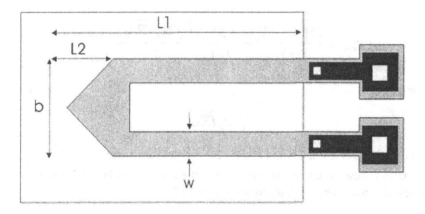

Figure 9.14: A typical piezoresistive cantilever layout.

The spring constant k, or the stiffness of the cantilever, can be expressed as

$$(9.8)$$

$$k = \frac{Yt^3 wb}{2b(L_1^3 - L_2^3) + 6wL_2^3} \qquad ,$$

where Y is the Young's modulus and t is the thickness of the cantilever. The fractional change in resistance is the sensitivity of the cantilever for a given displacement at the tip:

(9.9)

$$\frac{\Delta R}{R} = \frac{3\pi Y t b(L_1 + L_2)}{4\left[(L_1^3 - L_2^3)b + 3L_2^3 w\right]} y \qquad ,$$

The minimum detectable deflection (y_{min}) is the amount of deflection that produces a unity signal-to-noise ratio when a fractional change in resistance is combined with the noise figure for the system.

(9.10)

$$y_{min} = \frac{16\left[(L_1^3 - L_2^3)b + 3L_2^3 w\right]}{3\pi Y t b(L_1 + L_2)} \frac{\sqrt{2k_b TR \Delta f}}{V} \qquad ,$$

where

$$\sqrt{2k_b TR \Delta f}$$

is the thermal resistor noise and V is the bias to the wheatstone bridge.

Using these equations as a guideline, the cantilevers in the 2D array were designed to have the properties listed in Table 9.2.

Design	L1 (μm)	K (N/m)	ΔR/R per Angstrom
A	410	49.3	$6.25\ e^{-7}$
B	350	79.3	$8.78\ e^{-7}$
C	300	125.9	$1.23\ e^{-6}$
D	250	217.6	$1.83\ e^{-6}$
E	200	425	$3.03\ e^{-6}$

Table 9.2:   Design parameters of the cantilever in the 2D array.
All designs have L_2= 60 μm, W=40 μm, b=120 μm, and t=10 μm.

9.5.3 Processing

There are two important features that must be incorporated into the fabrication of the piezoresistive cantilevers for the 2D array. First we need probe tips with high aspect ratios so that we gain enough spacing between the probe array and the sample. Second, we need to integrate the TWV's into the cantilever fabrication process.

We have developed a method of sealing the vias during the processing with a standard thick spin-on resist. The electro-deposited resist could not be used because of materials compatibility issues in the processing equipment. The following lists details of the process, and the schematic diagrams corresponding to each process step are provided in Table 8.3.

The 2D arrays of cantilevers were fabricated on silicon-on-insulator (SOI) wafers which are polished on the front and back sides. The starting wafers consist of 20-μm-thick top layer silicon (lightly doped p-type) and 1.9-μm-thick buried oxide on top of a 500-μm-thick p-type <100> silicon substrate.

Step a: The tips were formed by first growing a 1-μm-thick thermal oxide which is patterned by photolithography. The oxide was wet etched in 6:1 BOE (buffered oxide etch). The oxide is used as the etch mask for the subsequent dry etching described in the previous section. After the tip formation, the oxidation sharpening was performed, and 100 nm of thermal oxide was regrown.

Step b: The global alignment marks (not shown in the schematic) were transferred into the oxide layer and will be used for overlay of subsequent masking layers. The piezoresistors were fabricated by ion implantation of boron through the 100 nm of oxide with a dose of $5 \times 10^{14}/$ cm^2 at 80 KeV. To form ohmic contacts to the resistors, the contact area was patterned and implanted with a second implant ($5 \times 10^{15}/$ cm^2, at 40 KeV). The dopants were activated by rapid thermal annealing.

Step c: The TWV's were defined by etching the via in the top silicon layer, and in the buried oxide. Both layers were etched by anisotropic RIE etching.

Step d: An 18–μm-thick photoresist (AZ 4620) was coated on the front and back sides of the wafer. Lithography of the TWV pattern was performed on the backside of the wafer using an aligner which is capable of aligning the front and the backside (Karl Suss). The TWV was etched for 3 hours (STS etcher). The etch was terminated when the front and the backside vias met each other. The resist coating on the front side protected the tips during the TWV etching.

Step e: Conformal thin film layers were deposited on both sides of the wafer: (1) 400-nm-thick LTO (low temperature oxide) was deposited to protect the tips. (2) 1-μm-thick LPCVD nitride was deposited to electrically isolate the vias. (3) 1-μm-thick LPCVD polysilicon was deposited as a sticking layer for the subsequent CVD tungsten deposition.

Step f: The contacts to the piezoresistors were patterned using the AZ 4620 photoresist. The polysilicon and nitride layers were etched in a plasma etcher (Drytek), and then the oxide layer was wet etched.

Step g: A 1-μm-thick tungsten film was deposited by chemical vapor deposition on both sides of the wafer The tungsten makes electrical contact to the heavily doped region of the piezoresistor. Aluminum was next sputter deposited only on the backside to allow wirebonding to the metal contacts.

Step h: While protecting the metallization on the back side with thick AZ4620 photoresist, the tungsten metallization on the front side was patterned and etched. The polysilicon/nitride/oxide stack was also etched using the same mask, exposing the top silicon layer of the SOI wafer.

Step i: With the front side of the wafer still protected with the thick photoresist, the aluminum/tungsten metallization and the polysilicon/nitride/oxide layers on the back side were patterned and etched. The resist was then stripped.

Step j: Thick photoresist films were spun on both sides of the wafer. The cantilevers were patterned on the front side of the wafer and the top silicon layer was etched using the STS anisotropic silicon etcher. The etch was terminated at the buried oxide. Then the resists were stripped, and forming gas anneal at 400 °C was performed to improve the contact resistance between the silicon and tungsten.

Step k: Thick photoresist was spun on to protect the front side of the wafer, and lithography of the cantilever release pattern was performed on the backside. The substrate silicon was etched in STS anisotropic silicon etcher for 2.5 hours. The etch was terminated when it reached the buried oxide. The buried oxide was wet etched in 6:1 BOE to free the cantilevers.

Step l: The resist on both sides of the wafer were etched in an oxygen plasma, completing the device fabrication.

The arrays must be cleaved from the wafer and wirebonded to a package before use. SEM micrographs of the completed devices are shown in Fig. 9.15.

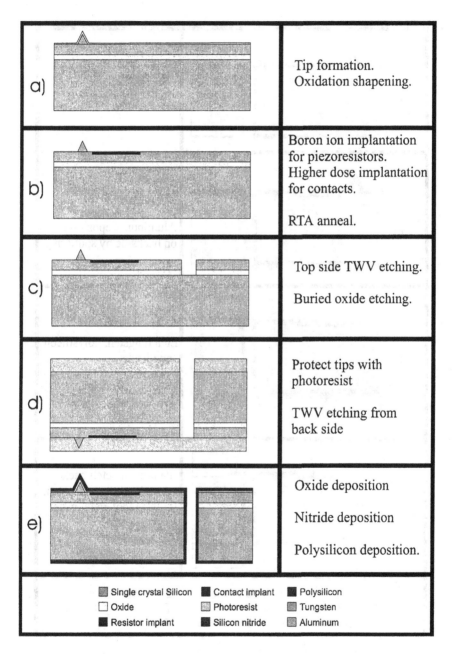

a)	Tip formation. Oxidation shapening.
b)	Boron ion implantation for piezoresistors. Higher dose implantation for contacts. RTA anneal.
c)	Top side TWV etching. Buried oxide etching.
d)	Protect tips with photoresist TWV etching from back side
e)	Oxide deposition Nitride deposition Polysilicon deposition.

Single crystal Silicon   Contact implant   Polysilicon
Oxide   Photoresist   Tungsten
Resistor implant   Silicon nitride   Aluminum

Table 9.3:   2D array process flow diagram

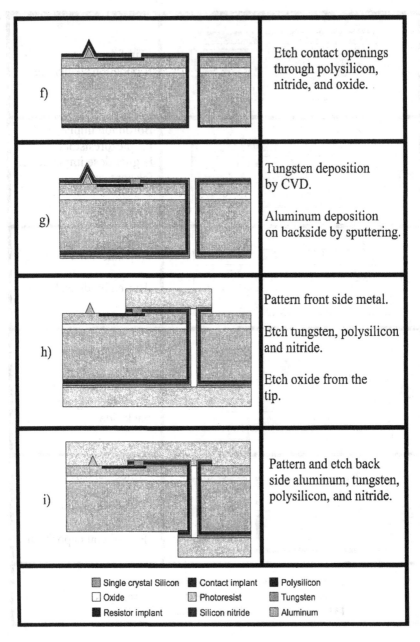

f)	Etch contact openings through polysilicon, nitride, and oxide.
g)	Tungsten deposition by CVD. Aluminum deposition on backside by sputtering.
h)	Pattern front side metal. Etch tungsten, polysilicon and nitride. Etch oxide from the tip.
i)	Pattern and etch back side aluminum, tungsten, polysilicon, and nitride.

Legend:
- Single crystal Silicon
- Oxide
- Resistor implant
- Contact implant
- Photoresist
- Silicon nitride
- Polysilicon
- Tungsten
- Aluminum

Table 8.3 continued

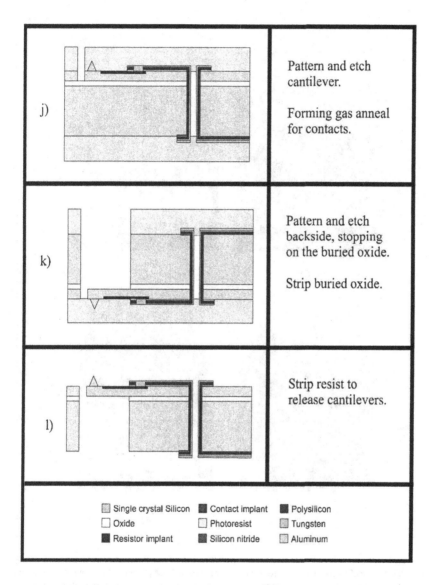

j)		Pattern and etch cantilever. Forming gas anneal for contacts.
k)		Pattern and etch backside, stopping on the buried oxide. Strip buried oxide.
l)		Strip resist to release cantilevers.

▦ Single crystal Silicon	■ Contact implant	■ Polysilicon
☐ Oxide	☐ Photoresist	▦ Tungsten
■ Resistor implant	■ Silicon nitride	▦ Aluminum

Table 8.3 continued

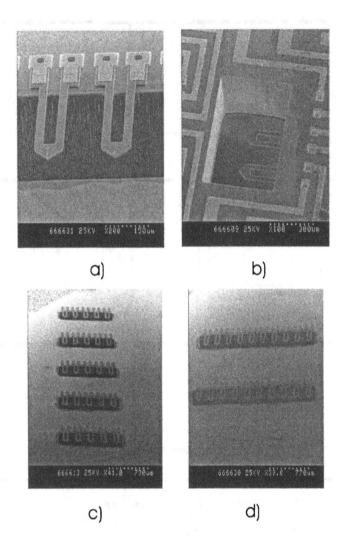

Figure 9.15: SEM micrographs of released 2D arrays of AFM probes. (a) View from the front side of the wafer showing tungsten interconnections from the TWV's to the piezoresistors. (b) View from the backside of the wafer showing the TWV's and tungsten/aluminum metal traces that connect to bonding pads (not shown). (c) 5×5 array. (d) 10×2 array of probes.

9.6   Imaging With the 2D Array

Completed cantilevers were 10 μm thick, 200 to 400 μm long, and had 7-μm-tall tips. An off-chip Wheatstone bridge circuit followed by a 10^5 gain stage was used to monitor changes in each cantilever's piezoresistance. We measured deflection sensitivities of 10^{-7} nm^{-1} to 5×10^{-7} nm^{-1} and minimum detectable deflections of 10-20 Å (10 Hz - 1 kHz bandwidth).

To demonstrate their functionality, a 2×4 cantilever array was used to image an arbitrary location on a grating which is shown in Fig. 9.16. After the bridge gains and offsets were individually tuned for each cantilever, the cantilever array was aligned to the sample. Each cantilever scanned 170 μm × 70 μm, while their signals were simultaneously collected by a computer. The entire scan was acquired in 140 sec. The fact that the sample was much larger than the cantilever array (1 cm^2 vs. 0.5 mm^2) demonstrates the utility of TWV integration.

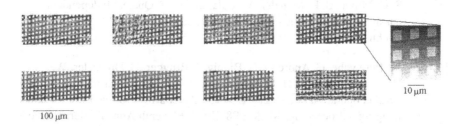

10 μm

100 μm

Figure 9.16:   Parallel image of a grating (10 μm pitch, 0.2 μm step) with a 2 x 4 array of piezoresistive cantilevers. The cantilever pitch is 250 μm horizontally, and 1500 μm vertically. Each cantilever sweeps out a 170 μm x 70 μm area by scanning at 560 μm/sec. The entire image is scanned in 140 sec. A scratch on the grating is visible across the top left two cantilevers.

The fabrication process is scalable to larger and denser arrays and can be integrated with other unique sensor topologies. Though specific parameters depend on the application, it is reasonable to consider arrays of thousands of probes scanning centimeter distances in tens of seconds. Increasing throughput through parallelism is particularly attractive for scanning probes because it allows larger sample regions and shorter acquisition times while maintaining functional versatility and force sensitivity.

9.7 References

[1] S. C. Minne, J. D. Adams, G. Yaralioglu, S. R. Manalis, A. Atalar, and C. F. Quate, "Centimeter-scale atomic force microscope imaging and lithography," Appl. Phys. Lett. **73**, 1741-1744 (1998).

[2] A. J. Peyton and V. Walsh, *Analog Electronics With Op Amps* (New York: Cambridge University Press, 1993).

[3] S. C. Minne, "Increasing the throughput of atomic force microscopy," Ph.D. Thesis, Stanford University, 1996.

[4] S. C. Minne S. R. Manalis, and C. F. Quate, "Parallel atomic force microscopy using cantilevers with integrated piezoresistive sensors and integrated piezoelectric actuators," Appl. Phys. Lett. **67**, 3918 (1995).

[5] S. C. Minne, S. R. Manalis, A. Atalar and C. F. Quate, "Independent parallel lithography using the atomic force microscope," J. Vac. Sci. Technol. B **14**, 2458 (1996).

[6] M. Lutwyche, C. Andreoli, G. Binnig, J. Brugger, U. Drechsler, W. Haeberle, H. Rohrer, H. Rothuizen, and P. Vettiger, "Microfabrication and parallel operation of 5×5 2D AFM cantilever arrays for data storage and imaging," Proceedings MEMS 98. IEEE. Eleventh Annual International Workshop on Micro Electro Mechanical Systems. (Cat. No.98CH36176), 8 (1998).

[7] *International Technology Roadmap for Semiconductors* (San Jose: Semiconductor Industry Association, 1997). Data reflect 1998 update to the roadmap.

[8] K. Bean, "Anisotropic etching of silicon", IEEE Trans.Electron Devices **ED-25**, 1187 (1978).

[9] O. Tabata, R. Asahi, H. Funabashi, K. Shimaoka, and S. Sugiyama, "Anisotripic etching of silicon in TMAH solutions," Sensors and Actuators A **34**, 51 (1992).

[10] E. M. Chow, H. T. Soh, A. Partridge, J. A. Harley, T. W. Kenny and C. F. Quate, "Fabrication of high density cantilever arrays and through-wafer connections," Proc. Solid State Sensors and Actuator Workshop, Hilton Head Island, South Carolina (6-11 June), 220, (1998).

[11]   C. Christensen, P. Kersten, S. Henke and S. Bouwstra, "Wafer through-hole interconnections with high vertical wiring densities," IEEE Transactions on Components, Packaging, and Manufacturing Technology **19**, 516 (1996).

[12]   P. Kersten, S. Bouwstra and J. W. Petersen, "Photolithography on micromachined 3D surfaces using electrodeposited photoresists," Sensors and Actuators A (Physical).**A51**, 51 (1995).

[13]   M. Tortonese, H. Yamada, R. C. Barrett, and C. F. Quate, "Atomic force microscopy using a piezoresistive cantilever," Proceedings of the 6th International Conference on Solid-State Sensors and Actuators (Transducers '91) **91**, 448 (1991).

[14]   M. Tortonese, "Force sensors for scanning probe microscopy," Ph.D. Thesis, Ginzton Laboratory Number 5098, Stanford University, June 1993.

Epilog

Patterning nanoscale features with a size less that 100 nm falls into the domain occupied by electron beam lithography (EBL). This is a mature technology with software in place, but it is restricted in several areas. It is not easy to pattern feature sizes below 30 nm. The proximity effect makes it difficult to place fine lines close to each other. With the large depth of focus it is difficult to register patterns with sub-100 nm features as required for overlay with multiple layers. The primary restriction is that of throughput. The EBL system is not often used for patterning multiple copies. It takes too long to write patterns over modest areas. For example, it may take four hours to fill one square centimeter with 50 nm dots spaced by 20 microns. Furthermore, the SEM is not readily configured for arrays which the barrier for reaching high throughput with this technology.

Scanning probe lithography (SPL), with the attributes discussed in this book, is an alternative to EBL. There is no proximity effect and the lines can be less than 10 nm in width when carbon nanotubes are used as tips. With piezoelectric actuators we can scan the arrays at speeds of 10 mm/sec. With this combination SPL becomes a contender for advanced lithography systems.

The principal challenge for SPL is throughput. We envision very large array as the solution. We cite two examples where arrays are now operational. Vettiger at IBM in Zurich reports on an array of 32×32 cantilevers.[1] Their chip, which is 3 mm on a side, is designed for digital storage, but these can be easily converted for lithography. When the tips are scanned at 5 mm/sec it will take 10 secs to pattern this area with 100 nm pixels. This translates to a patterning rate of 0.5 mm^2/sec which is much slower than the rate of 100 mm^2/sec achieved with current systems. To reach the higher rates we might turn to the large array as announced by Daewoo Electronics.[2] That group has built an array of micromirrors designed for projection display. Each micromirror is mounted on a cantilever that is bent to tilt the micromirror. The cantilevers are actuated with piezoelectric films deposited on the individual cantilevers. The mirrors can be easily replaced with integrated tips to

1. M. Despont, J. Brugger, U. Drechsler, U. Durig, W. Haberle, M. Lutwyche, H Rothuizen, R. Widmer, G. K. Binnig, and P. Vettiger, "Microfabrication and Testing of 32x32 Cantilever Array Chip for AFM Storage," p. 564-569, Proceeding of 12th IEEE Int'l Micro Electro Mechanical Systems Conference, "MEMS '99," Orlando, Florida, Jan 17-21, 1999.
2. S. G. Kim and K.-H. Hwang, "Thin-Film Micrromirror Array," Information Display **15**, 30-34, (April/May 1999).

adapt these arrays for SPL.[1] Their array is very large, 780×1024 elements, covering a chip 2.5 inches on a side with good uniformity over the entire area. The array with 800,000 elements covers 40 cm^2. With a scanning speed of 4 mm/sec this area can be scanned in four seconds, which translates to a throughput of 100 mm^2/sec. When the pixel size is reduced to 30 nm the scanning speed must be increased to 12 mm/sec to maintain this throughput. The potential for writing sub-30-nm features means that this technology can be used for several generations beyond 100 nm.

---

1. www.tmadisplay.com

List of Publications

The body of this book is based on research performed at Stanford University, parts of which have been previously published in the following:

[1] K. S. Wilder, "Maskless lithography using scanning probes," Ph.D. Thesis, Department of Applied Physics, Stanford University (Aug 1999).

[2] H. T. Soh, "Advancements in scanning probe lithography and nanostructure fabrication," Ph.D. Thesis, Department of Electrical Engineering, Stanford University (Mar 1999).

[3] H. T. Soh, A. F. Morpurgo, J. Kong, C. M. Marcus, C. F. Quate, and H. Dai, "Integrated nanotube circuits: Controlled growth and ohmic contacting of single-walled carbon nanotubes," Appl. Phys. Lett. (1999).

[4] K. Wilder and C. F. Quate, "Scanning probe lithography using a cantilever with integrated transistor for on-chip control of the exposing current," J. Vac. Sci. Technol. B (1999).

[5] K. Wilder, H. T. Soh, A. Atalar, and C. F. Quate, "Nanometer-scale patterning and individual current-controlled lithography using multiple scanning probes," Rev. Sci. Instrum. **70**, 2822-2827 (1999).

[6] J. D. Adams, S. C. Minne, S. R. Manalis, K. Wilder, G. Yaralioglu, D. Adderton, and C. F. Quate, "High throughput, high resolution scanning probe microscopy," Future Fab International **6**, 259-263 (1999).

[7] H. T. Soh, C. P. Yue, A. M. McCarthy, C. Ryu, T. H. Lee, and C. F. Quate, "Ultra-low resistance, through-wafer via (TWV) technology and its applications in three dimensional structures on silicon," Jap. J. Appl. Phys. (1999).

[8] H. T. Soh, J. Kong, A. M. Cassell, C. F. Quate, and H. Dai, "Synthesis of individual single-walled carbon nanotubes on patterned silicon wafers," Nature **6705**, 878 (1998).

[9] K. Wilder, C. F. Quate, B. Singh, and D. F. Kyser, "Electron beam and scanning probe lithography: A comparison," J. Vac. Sci. Technol. B **16**, 3864-3873 (1998).

[10] K. Wilder, C. F. Quate, D. Adderton, R. Bernstein, and V. Elings, "Noncontact

nanolithography using the atomic force microscope," Appl. Phys. Lett. **73**, 2527-2530 (1998).

[11] H. T. Soh, C. P. Yue, A. M. McCarthy, C. Ryu, T. H. Lee, and C. F. Quate, "Ultra-low resistance, through-wafer via (TWV) technology and its applications in three dimensional structures on silicon," Proc. International Conference on Solid State Devices and Materials (SSDM '98), 284 (1998).

[12] E. M. Chow, H. T. Soh, A. Partridge, J. A. Harley, T. W. Kenny, C. F. Quate, S. Abdollahi-Alibeik, J. McVittie, and A. M. McCarthy, "Fabrication of high-density cantilever arrays and through-wafer interconnects," Proceedings of 6th Sensors and Actuators Workshop, 220 (1998).

[13] G. Percin, H. T. Soh, and B. T. Khuri-Yakub, "Resist deposition without spinning by using novel inkjet technology and direct lithography for MEMS," Proceedings of 1998 SPIE's 23rd Annual International Symposium on Microlithography, 3333 pt. 1-2, 1382-9 (1998).

[14] K. Wilder, H. T. Soh, A. Atalar, and C. F. Quate, "Hybrid atomic force / scanning tunneling lithography," J. Vac. Sci. Technol. B **15**, 5, 1811-1817 (1997).

[15] H. T. Soh, K. Wilder, A. Atalar, and C. F. Quate, "Fabrication of 100 nm pMOSFETs with hybrid AFM / STM lithography," *Proceedings of the 1997 Symposium on VLSI Technology* (Japan Society of Applied Physics, IEEE Cat. No. 97CH36114, 1997), 129-130.

[16] H. T. Soh, K. Wilder, and C. F. Quate "Scanning probe lithography for electron device fabrication," Proceedings of the 1997 Samsung Humantech Thesis Competition (1997).

[17] K. Wilder, H. T. Soh, S. C. Minne, S. R. Manalis, and C. F. Quate, "Cantilever arrays for lithography," Naval Research Reviews **XLIX**, 1, 35-48 (1997).

[18] K. Wilder, C. F. Quate, B. Singh, R. Alvis, and W. H. Arnold, "Atomic force microscopy for cross section inspection and metrology," J. Vac. Sci. Technol. B **14**, 6, 4004-4008 (1996).

[19] K. Wilder, B. Singh, and W. H. Arnold, "Scanning probe applications in the semiconductor industry," Future Fab International **1**, 256-260 (1996).

[20] S. W. Park, H. T. Soh, C. F. Quate, and S.-I. Park, "Nanometer scale lithography at high scanning speeds with the atomic force microscope using spin on glass," Appl. Phys. Lett. **67**, 2415 (1995).

[21] S. C. Minne, H. T. Soh, Ph. Flueckiger, and C. F. Quate, "Fabrication of 0.1 um

metal oxide semiconductor field-effect transistors with the atomic force microscope," Appl. Phys. Lett. **66**, 703 (1995).

[22] S. C. Minne, Ph. Flueckiger, H. T. Soh, and C. F. Quate, "Atomic force microscope lithography using amorphous silicon as a resist and advances in parallel operation," J. Vac. Sci, Technol. B **13**, 1380 (1995).

Index